ALFRED SCHREIBER

STIRLING POLYNOMIALS
IN SEVERAL INDETERMINATES

ALFRED SCHREIBER

STIRLING POLYNOMIALS
IN SEVERAL INDETERMINATES

Logos Verlag Berlin

λογος

Bibliographic information published by the Deutsche Nationalbibliothek

The Deutsche Nationalbibliothek lists this publication in the Deutsche Nationalbibliografie; detailed bibliographic data are available on the Internet at http://dnb.d-nb.de.

ISBN 978-3-8325-5250-3

Logos Verlag Berlin GmbH
Georg-Knorr-Str. 4, Geb. 10
D-12681 Berlin
Tel.: +49 (0)30 42 85 10 90
Fax: +49 (0)30 42 85 10 92
INTERNET: http://www.logos-verlag.com

PREFACE

The two chapters of this little book originally were research papers on ›multivariate Stirling polynomials‹, the latter being a collective term for the partial Bell polynomials and their orthogonal companions. As far as this topic needs an explanation, I would like to briefly describe the starting point and the main ideas of my work.

Suppose φ is an analytic function with $\varphi(0) = 0$ and k is any positive integer. We consider the problem of expanding $\varphi(x)^k$ into a Taylor series at the point 0. In the case $k = 1$ this amounts to determining the Taylor coefficients $\varphi_0(= 0)$, $\varphi_1, \varphi_2, \varphi_3, \ldots$ of φ. In the general case, the coefficient $[x^n/n!]\varphi(x)^k$ can be represented by a surprisingly elegant expression, namely $k!B_{n,k}(\varphi_1, \ldots, \varphi_{n-k+1})$. Here $B_{n,k}(\varphi_1, \ldots, \varphi_{n-k+1})$ are those formerly so-called (partial) exponential polynomials in $\varphi_1, \ldots, \varphi_{n-k+1}$, which, following a suggestion from Riordan, are now commonly called (partial) Bell polynomials. They can be viewed as partition polynomials, each monomial of which describes a certain partition of a set of n objects into k blocks. Therefore, the sum of the coefficients is just equal to the number of all such partitions. More precisely, we have $B_{n,k}(1, \ldots, 1) = s_2(n, k)$, where $s_2(n, k)$ denotes the Stirling numbers of the second kind.

Chapter I deals with the problem of finding a family of polynomials $A_{n,k}$, which are related to the $B_{n,k}$ and, in an analogous manner, yield the signed Stirling numbers of the first kind: $A_{n,k}(1, \ldots, 1) = s_1(n, k)$. There are several approaches here, all of which have in common that they are based on the inversion of the underlying function φ. So, for example, if in our consideration above we replace φ by its compositional inverse $\overline{\varphi}$, the Taylor coefficient of $\overline{\varphi}(x)^k/k!$ becomes $B_{n,k}(\overline{\varphi}_1, \ldots, \overline{\varphi}_{n-k+1})$. The latter now has to be converted into an expression that solely depends on $\varphi_1, \varphi_2, \varphi_3, \ldots$ Such an expression actually exists in the form of a Laurent polynomial $A_{n,k}(\varphi_1, \ldots, \varphi_{n-k+1})$ which, in addition to the "variable" φ_1, also contains its reciprocal φ_1^{-1}. This sufficiently justifies $A_{n,k}$ and $B_{n,k}$ to be called multivariate Stirling polynomials of the first and second kind, respectively. This is reaffirmed by the fundamental relationships between them that will be set up in the course of Chapter I. Some of these are well-known identities for Stirling numbers, which are now raised to the more complex level of polynomial expressions, such as the orthogonality relation, the mutual Schlömilch type representations, and, as a high-light in Chapter II, the famous reciprocity (or, duality) law.

In Chapter II, the methods and results of Chapter I are refined and expanded in various ways. First, the axiomatic framework is now replaced with an algebraic standard model, in which functions are essentially represented by formal power series. This seems appropriate insofar as the Taylor coefficients of a function can take on the role of the indeterminates in a polynomial. Second, the focus is now on a larger class of polynomials. On the one hand, these are certain linear combinations of Bell polynomials (derived from the Faà di Bruno formula) and, on the other hand, "B-representable" expressions that result from a kind of plethystic substitution into Bell polynomials. Since in many cases this substitution corresponds to the composition of functions, the composition rules set out in Section II.4 are of particular methodological importance. The second of these rules, which goes back to Eri Jabotinsky, often makes proofs easier.

On this basis, a number of new results are presented, which concern different types of inverse relationships; for example, the use of multivariable Lah polynomials for characterizing self-orthogonal families of polynomials that can be represented by Bell polynomials; or, the introduction of "generalized Lagrange inversion polynomials" that invert functions characterized in a specific way by sequences of constants.

The versions of the two chapters presented below are revisions of the original preprints, which in many passages may differ considerably from the corresponding articles that were later published in journals [84, 86]. Nonetheless, I have moved away from the tempting idea of reprocessing all of the material and of reorganizing it in a new monograph. This would of course have offered the opportunity to harmonize the parts with one another, to avoid repetitions and update one or the other detail. On the whole, however, the risk seemed too great to me that I would eventually no longer be able to complete this newly opened work process (not least in view of my age). Every mathematician all too well knows the dilemma that arises when a seemingly harmless detail in his or her work begins to create new challenges that one cannot resist. I hope that the ›Notes and supplements‹ attached to both chapters compensate for this deficiency to some extent.

Alfred Schreiber
Dresden, January 27, 2021

CONTENT

One must always invert.

CARL G. J. JACOBI

I

MULTIVARIATE STIRLING POLYNOMIALS

1 Introduction

1.1 Background and problem

It is well-known that a close connection exists between iterated differentiation and Stirling numbers (see, e. g., [44, 77, 101]). Let $s_1(n,k)$ denote the signed Stirling numbers of the first kind, $s_2(n,k)$ the Stirling numbers of the second kind, and D the operator d/dx. Then, for all positive integers n, the nth iterate $(xD)^n$ can be expanded into the sum

$$(xD)^n = \sum_{k=1}^{n} s_2(n,k) x^k D^k. \tag{1.1}$$

An expansion in the reverse direction is also known to be valid (see, e. g., [44, p. 197] or [77, p. 45]):

$$D^n = x^{-n} \sum_{k=1}^{n} s_1(n,k)(xD)^k. \tag{1.2}$$

Let us first look at Eq. (1.1). The occurrence of the Stirling numbers can be explained combinatorially as follows. Observing

$$(xD)^n f(x) = D^n (f \circ \exp)(\log x)$$

we can use the classical higher-order chain rule (named after Faà di Bruno; cf. [42, 44], [51, pp. 52, 481]) to calculate the nth derivative of the composite function $f \circ g$:

$$(f \circ g)^{(n)}(x) = \sum_{k=1}^{n} B_{n,k}(g'(x), \ldots, g^{(n-k+1)}(x)) \cdot f^{(k)}(g(x)), \tag{1.3}$$

where $B_{n,k} \in \mathbb{Z}[X_1, \ldots, X_{n-k+1}]$, $1 \le k \le n$, is the (partial) exponential Bell polynomial

$$B_{n,k}(X_1, \ldots, X_{n-k+1}) = \sum_{r_1, r_2, \ldots} \frac{n!}{r_1! r_2! \ldots (1!)^{r_1} (2!)^{r_2} \ldots} X_1^{r_1} X_2^{r_2} \ldots \tag{1.4}$$

the sum to be taken over all non-negative integers $r_1, r_2, \ldots, r_{n-k+1}$ such that $r_1 + r_2 + \ldots + r_{n-k+1} = k$ and $r_1 + 2r_2 + \ldots + (n-k+1)r_{n-k+1} = n$. The coefficient in $B_{n,k}$ counts the partitions of n distinct objects into k blocks (subsets) with r_j blocks containing exactly j objects ($1 \leq j \leq n-k+1$). Therefore, the sum of these coefficients is equal to the number $s_2(n,k)$ of all such partitions. So we have $B_{n,k}(x, \ldots, x) = s_2(n,k)x^k$. Evaluating $(f \circ \exp)^{(n)}(\log x)$ by Eq. (1.3) then immediately gives the right-hand side of Eq. (1.1).

Question. Can also Eq. (1.2) be interpreted in this way by substituting jth derivatives in place of the indeterminates X_j of some polynomial $S_{n,k} \in \mathbb{Z}[X_1, \ldots, X_{n-k+1}]$, the coefficients of which add up to $s_1(n,k)$?

The main purpose of the present chapter is to give a positive and comprehensive answer to this question including recurrences, a detailed study of the inverse relationship between the polynomial families $B_{n,k}$ and $S_{n,k}$, as well as fully explicit formulas (with some applications to Stirling numbers and Lagrange inversion).

The issue turns out to be closely related to the problem of generalizing Eq. (1.1), that is, finding an expansion for the operator $(\theta D)^n$ ($n \geq 1$, θ a function of x). Note that, in the case of scalar functions, $(\theta D)f$ is the *Lie derivative* of f with respect to θ. Several authors have dealt with this problem. In [21] and [69] a polynomial family $C_{n,k} \in \mathbb{Z}[X_0, X_1, \ldots, X_{n-k}]$ has been defined[1] by differential recurrences and shown to comply with $(\theta D)^n = \sum_{k=1}^{n} C_{n,k}(\theta, \theta', \ldots, \theta^{(n-k)})D^k$. Comtet [21] has tabulated $C_{n,k}$ up to $n = 7$ and stated that $C_{n,k}(x, \ldots, x) = c(n,k)x^n$, where $c(n,k) := |s_1(n,k)|$ denotes the signless Stirling numbers of the first kind ('cycle numbers' according to the terminology in [50]). Since however all coefficients of $C_{n,k}$ are positive, $C_{n,k}$ does not appear to be a suitable companion for $B_{n,k}$ with regard to the desired inversion law.

Todorov [99, 100] has studied the above Lie derivation with respect to a function θ of the special form $\theta(x) = 1/\varphi'(x)$, $\varphi'(x) \neq 0$. His main results in [99] ensure the existence of $S_{n,k} \in \mathbb{Z}[X_1, \ldots, X_{n-k+1}]$ such that

$$\left(\varphi'(x)^{-1}D\right)^n f(x) = \sum_{k=1}^{n} A_{n,k}(\varphi'(x), \ldots, \varphi^{(n-k+1)}(x)) \cdot f^{(k)}(x), \quad (1.5)$$

---

[1] Here and in Chapter II we write $C_{n,k}$ instead of Comtet's $A_{n,k}$ (cf. [21]) in order to avoid misunderstandings. Note that in both chapters of this book $A_{n,k}$ is exclusively used to denote the 'Lie coefficients' according to Todorov (see Eq. (1.5) below).

where $A_{n,k} := X_1^{-(2n-1)} S_{n,k}$. While differential recurrences for $A_{n,k}$ can readily be derived from Eq. (1.5) (cf. [99, Equation (27)] or a slightly modified version in [100, Theorem 2]), a simple representation for $S_{n,k}$ — as is Eq. (1.4) for $B_{n,k}$ — was still lacking up to now. Todorov [99, p. 224] erroneously believed that the somewhat cumbersome 'explicit' expression in [21] for the coefficients of $C_{n,k}$ would directly yield the coefficients of $S_{n,k}$. Also the determinantal form presented in [99, Theorem 6] for $(D/\varphi')^n$ (and thus also for $S_{n,k}$) may only in a modest sense be regarded as explicit.

Nevertheless, Todorov's choice ($\theta = 1/\varphi'$) eventually proves to be a crucial idea. Among other things, it reveals that $A_{n,k}$ (and thus $S_{n,k}$) is connected with the classical Lagrange problem of computing the compositional inverse $\overline{f}$ of a given series $f(x) = \sum_{n \geq 1} (f_n/n!) x^n$, $f_1 \neq 0$. As we shall see later, the Taylor coefficients $\overline{f}_n$ of $\overline{f}(x)$ can be expressed simply by applying $A_{n,1}$ to the coefficients of f as follows:

$$\overline{f}_n = A_{n,1}(f_1, \ldots, f_n). \tag{1.6}$$

On the other hand, Comtet [22] found an inversion formula that expresses $\overline{f}_n$ in terms of the (partial) exponential Bell polynomials:

$$\overline{f}_n = \sum_{k=0}^{n-1} (-1)^k f_1^{-n-k} B_{n+k-1,k}(0, f_2, \ldots, f_n). \tag{1.7}$$

This result has been shown by Haiman and Schmitt [33, 81] to provide essentially both a combinatorial representation and a cancellation-free computation of the antipode on a Faà di Bruno Hopf algebra (a topic that has received a lot of attention in quantum field theory due to its application to renormalization; cf. [55, 20, 28]). Combining Eq. (1.6) with Eq. (1.7) we obtain an expression for $A_{n,1}$ in terms of the Bell polynomials. This suggests looking for a similar representation for the whole family $A_{n,k}$, $1 \leq k \leq n$. As a main result (Theorem 6.1), we shall prove the following substantially extended version of Eq. (1.6) & Eq. (1.7):

$$A_{n,k} = \sum_{r=k-1}^{n-1} (-1)^{n-1-r} \binom{2n-2-r}{k-1} X_1^{-(2n-1)+r} \widetilde{B}_{2n-1-k-r, n-1-r}. \tag{1.8}$$

The tilde over B indicates that X_1 has been replaced by 0. From Eq. (1.8) we eventually get the desired explicit standard representation for $A_{n,k}$ that corresponds to the one for $B_{n,k}$ given in Eq. (1.4).

Equation Eq. (1.8) states a somewhat intricate relationship between the families $A_{n,k}$ and $B_{n,k}$. A simpler connection of both expressions is the following basic inversion law, which generalizes the orthogonality of the Stirling numbers (cf. Section 5):

$$\sum_{j=k}^{n} A_{n,j} B_{j,k} = \delta_{nk} \quad (1 \le k \le n), \tag{1.9}$$

where $\delta_{nn} = 1$, $\delta_{nk} = 0$ if $n \ne k$ (Kronecker symbol).

1.2 Terminology and notation

Considering Eq. (1.9) and the fact that the sum of the coefficients of $A_{n,k}$ and of $B_{n,k}$ are equal to $s_1(n,k)$ and to $s_2(n,k)$, respectively, it may be justified to call $A_{n,k}$ and $B_{n,k}$ *multivariate Stirling polynomials of the first and second kind*. There should be no risk of confusing them with polynomials *in one variable* like those introduced and named after Stirling by Nielsen [73, 74], neither with the closely related 'Stirling polynomials' $f_k(n) := s_2(n+k, n)$ and $g_k(n) := c(n, n-k)$ Gessel and Stanley [31] have investigated as functions of $n \in \mathbb{Z}$.

A sequence $r_1, r_2, r_3, \ldots$ of non-negative integers is said to be an (n, k)-*partition type*, $0 \le k \le n$, if $r_1 + r_2 + r_3 + \ldots = k$ and $r_1 + 2r_2 + 3r_3 + \ldots = n$. The set of all (n, k)-partition types is denoted by $\mathbb{P}(n, k)$; we write $\mathbb{P}$ for the union of all $\mathbb{P}(n, k)$. In the degenerate case $(k = 0)$ set $\mathbb{P}(n, 0) = \emptyset$, if $n > 0$, and $\mathbb{P}(0, 0) = \{0\}$ otherwise. Let $k \ge 1$. Since $n - k + 1$ is the greatest j such that $r_j > 0$, we often write (n, k)-partition types as ordered $(n-k+1)$-tuples $(r_1, \ldots, r_{n-k+1})$.

The polynomials to be considered in the sequel have the form

$$P_\pi = \sum \pi(r_1, r_2, \ldots) X_1^{r_1} X_2^{r_2} \ldots,$$

where the sum ranges over all elements $(r_1, r_2, \ldots)$ of a full set $\mathbb{P}(n, k)$. As a consequence, P_π is homogeneous of degree k and isobaric of degree n. The coefficients of P_π may be regarded as values of a map $\pi : \mathbb{P} \longrightarrow \mathbb{Z}$ defined by some combinatorially meaningful expression, at least in typical cases like the following:

$$\omega(r_1, r_2, \ldots) := \frac{(r_1 + 2r_2 + \ldots)!}{r_1! \, r_2! \cdot \ldots} \qquad \textit{order function} \text{ (Lah)} \tag{1.10}$$

$$\zeta(r_1, r_2, \ldots) := \frac{\omega(r_1, r_2, \ldots)}{1^{r_1} \, 2^{r_2} \cdot \ldots} \qquad \text{\textit{cycle function}} \text{ (Cauchy)} \qquad (1.11)$$

$$\beta(r_1, r_2, \ldots) := \frac{\omega(r_1, r_2, \ldots)}{(1!)^{r_1} \, (2!)^{r_2} \cdot \ldots} \qquad \text{\textit{subset function}} \text{ (Faà di Bruno)} \qquad (1.12)$$

These coefficients count the number of ways a set can be partitioned into non-empty blocks according to a given partition type, that is, r_j denotes the number of blocks containing j elements ($j = 1, 2, \ldots$). The result depends on the meaning of 'block': linearly ordered subset (ω), cyclic order (ζ), or unordered subset (β).

It should be noticed that the corresponding polynomials $P_\omega, P_\zeta, P_\beta (= B_{n,k})$ are closely related to well-known combinatorial number-families:

$$P_\omega(1, \ldots, 1) = l^+(n, k), \text{ unsigned Lah numbers [57, 77]}$$

$$P_\zeta(1, \ldots, 1) = c(n, k) = \begin{bmatrix} n \\ k \end{bmatrix}, \text{ unsigned Stirling numbers of the 1st kind}$$

$$P_\beta(1, \ldots, 1) = s_2(n, k) = \begin{Bmatrix} n \\ k \end{Bmatrix}, \text{ Stirling numbers of the 2nd kind.}$$

1.3 Overview

This chapter is organized as follows: In Section 2 a general setting is sketched that allows functions and derivations to be treated algebraically. Section 3 contains a study of the iterated Lie operator $D(\varphi)^{-1}D$. An expansion formula for $(D(\varphi)^{-1}D)^n$ is established together with a differential recurrence for the resulting Lie coefficients $A_{n,k}$. Doing the same with respect to the inverse function $\overline{\varphi}$ will yield, conversely, D^n expanded and $B_{n,k}$ as the corresponding Lie coefficients. A by-product of Section 3 is Faà di Bruno's formula and its applications to the partial Bell polynomials $B_{n,k}$ to be briefly summarized in Section 4. These basic facts then lead to both inversion and recurrence relations, which we shall demonstrate and discuss in Section 5. The main task in Section 6 is to find an explicit polynomial expression for $S_{n,k}$. This is eventually achieved by means of Eq. (1.8), a proof of which makes up a central part of the section. In Section 7 we give some applications to the Lagrange inversion problem and to exponential generating functions.

2 Function algebra with derivation

2.1 Basic notions

Menger [64] has introduced the notion of a 'tri-operational algebra' of functions, which in the sequel (since 1960) stimulated to a great extent studies of generalized function algebras, e. g., [26, 65, 89, 90, 96]. In what follows I will propose a variant of Menger's original system tailored to our specific purposes of treating functions and their derivatives in a purely algebraic way.

Let $(\mathcal{F}, +, \cdot)$ be a non-trivial commutative ring of characteristic zero, 0 and 1 its identity elements with respect to addition and multiplication. We will think of the elements of $\mathcal{F}$ as 'functions (of one variable)' and therefore assume that $\mathcal{F}$ has a third binary operation $\circ$ (called *composition*) together with an identity element ι such that the following axioms are satisfied:

(F1) $f \circ (g \circ h) = (f \circ g) \circ h$

(F2) $(f + g) \circ h = (f \circ h) + (g \circ h)$

(F3) $(f \cdot g) \circ h = (f \circ h) \cdot (g \circ h)$

(F4) $f \circ \iota = \iota \circ f = f$

(F5) $1 \circ 0 = 1$

(F4) is assumed to be valid for all $f \in \mathcal{F}$; hence ι is unique. Let f be any element of $\mathcal{F}$. From (F2) we conclude $0 \circ f = 0$; so we get $\iota \neq 0$ (by (F4)) and $\iota \neq 1$ (by (F5)). (F2) furthermore implies $(-f) \circ g = -(f \circ g)$.

The least subring of $\mathcal{F}$ containing 1 will in the following conveniently be identified with $\mathbb{Z}$. (F5) then extends to the integers, that is, $n \circ 0 = n$ holds for all $n \in \mathbb{Z}$.

Given a *unit* f in $\mathcal{F}$ (i. e., f is an element invertible with respect to multiplication), we write f^{-1} (or $1/f$) for the inverse (henceforth called *reciprocal*) of f.

Remark 2.1. It must be emphasized that $\circ$ has to be understood as a *partial* operation (of course, $\iota^{-1} \circ 0$ is not defined). We therefore assign truth values to formulas, especially to our postulates (F1–3), whenever the terms involved are meaningful.

Let $f, g \in \mathcal{F}$ be functions such that $f \circ g = g \circ f = \iota$. Then g is called the *compositional inverse* of f, and vice versa. It is unique and will be denoted by $\overline{f}$. The following is obvious: $\overline{\iota} = \iota$, $\overline{\overline{f}} = f$, and $\overline{f \circ g} = \overline{g} \circ \overline{f}$.

Definition 2.1. Suppose $(\mathcal{F}, +, \cdot, \circ)$ satisfies (F1–5). We then call a mapping $D : \mathcal{F} \longrightarrow \mathcal{F}$ *derivation on $\mathcal{F}$*, and $(\mathcal{F}, +, \cdot, \circ, D)$ a *function algebra with derivation*, if D meets the following conditions:

- (D1) $D(f + g) = D(f) + D(g)$
- (D2) $D(f \cdot g) = D(f) \cdot g + f \cdot D(g)$
- (D3) $D(f \circ g) = (D(f) \circ g) \cdot D(g)$
- (D4) $D(\iota) = 1$
- (D5) $D(f) = 0 \Longrightarrow f \circ 0 = f$

The classical derivation rules (D1), (D2) make $\mathcal{F}$ into a differential ring. Some simple facts are immediate: $D(0) = D(1) = 0$, $D(m \cdot f) = m \cdot D(f)$ for all $m \in \mathbb{Z}$. By an inductive argument the product rule (D2) can be generalized:

$$D(f_1 \cdots f_n) = \sum_{k=1}^{n} f_1 \cdots f_{k-1} \cdot D(f_k) \cdot f_{k+1} \cdots f_n. \tag{2.1}$$

By putting $f_i = f$, $1 \leq i \leq n$, Eq. (2.1) becomes $D(f^n) = n f^{n-1} D(f)$. If f is a unit, this holds also for $n \leq 0$. As usual, f^m for $m < 0$ is defined by $(f^{-1})^{-m}$.

(D4) prevents D from operating trivially. In the case of a field $\mathcal{F}$, (D4) can be weakend to $D(f) \neq 0$ (for some $f \in \mathcal{F}$), since the chain rule (D3) then gives $D(f) = D(f \circ \iota) = D(f) \cdot D(\iota)$.

Applying (D3) and (D4) to $D(f \circ \overline{f})$ we obtain the inversion rule

$$D(\overline{f}) = \frac{1}{D(f) \circ \overline{f}}. \tag{2.2}$$

In a differential ring, it is customary to define the subring $\mathcal{K}$ of constants as the kernel of the additive homomorphism D, that is,

$$\mathcal{K} := \{ f \in \mathcal{F} \mid D(f) = 0 \}.$$

We have $\mathbb{Z} \subseteq \mathcal{K}$. Constants behave as one would expect.

Proposition 2.1. $c \in \mathcal{K} \iff c \circ f = c$ *for all $f \in \mathcal{F}$.*

Proof. $\Rightarrow$: Suppose $D(c) = 0$. Then, for any $f \in \mathcal{F}$ we have by (F1) and (D5): $c \circ f = (c \circ 0) \circ f = c \circ (0 \circ f) = c \circ 0 = c. - \Leftarrow$: Set $f = 0$ and apply the chain rule (D3). $\diamond$

If $f \circ 0$ exists for $f \in \mathcal{F}$, then it is obviously a constant.

Examples 2.1 (Function algebras with derivation). In each case below, the 'functions' carry some argument X (indeterminate, variable) that can be substituted in the usual sense: $f(g(X)) =: (f \circ g)(X)$.

1. The rational function field $\mathbb{Q}(X)$ together with the algebraically defined derivation $R \longmapsto R'$, $R \in \mathbb{Q}(X)$.
2. The ring $\mathbb{R}[[X]]$ of all power series with formal differentiation.
3. The ring of real-valued C^∞ functions on an open real interval with the ordinary differential operator d/dx.
4. The field of all meromorphic functions on a given region in $\mathbb{C}$ with complex differentiation.

Given a polynomial $P \in \mathcal{K}[X_1, \ldots, X_n]$ and functions $f_1, \ldots, f_n$, we denote by $P(f_1, \ldots, f_n)$ the function obtained by substituting f_i in place of X_i, $1 \le i \le n$. Recalling the algebraic definition of $\partial/\partial X_i$, we readily obtain by Eq. (2.1) the generalized chain rule:

$$D(P(f_1,\ldots,f_n)) = \sum_{k=1}^{n} \frac{\partial P}{\partial X_k}(f_1, \ldots, f_n) \cdot D(f_k). \qquad (2.3)$$

In the case $n = 1$ this becomes $D(P(f)) = P'(f) \cdot D(f)$. Thus, D restricted to polynomial functions turns out to act like the ordinary differential operator: $D(P(\iota)) = P'(\iota)$. Equation (2.3) also applies to the case that P is a rational function from $\mathcal{K}(X_1, \ldots, X_n)$.

Notation. Suppose that φ is a fixed function, and let D^i denote the ith iterate of D. In the following we will abbreviate $P(D(\varphi), D^2(\varphi), \ldots, D^n(\varphi))$ to P^φ. We denote by $\mathcal{P}_n(\varphi)$ the set of all P^φ and by $\mathcal{R}_n(\varphi)$ the set of all rational expressions P^φ/Q^φ, where $P, Q \in \mathcal{K}[X_1, \ldots, X_n]$.

Remark 2.2. Obviously $(P + Q)^\varphi = P^\varphi + Q^\varphi$ and $(P \cdot Q)^\varphi = P^\varphi \cdot Q^\varphi$. For a homogeneous polynomial P of degree k and arbitrary $a, b \in \mathcal{K}$ we have $P^{a\varphi+b} = a^k P^\varphi$.

Another formula that is known from calculus and that holds in the differential ring $(\mathcal{F}, D)$ is the general Leibniz rule, which yields an explicit expression for the higher derivatives of a product:

$$D^s(f_1 \cdots f_n) = \sum_{\substack{j_1+\cdots+j_n=s \\ j_1,\ldots,j_n \ge 0}} \frac{s!}{j_1! \cdots j_n!} \, D^{j_1}(f_1) \cdots D^{j_n}(f_n). \qquad (2.4)$$

We note the special case $D^s(\iota^n) = (n)_s \cdot \iota^{n-s}$, if $s \leq n$, and $D^s(\iota^n) = 0$ otherwise; the falling power $(n)_s$ is defined by $n(n-1)\cdots(n-s+1)$, $(n)_0 = 1$.

Convention. *Throughout the remainder of this chapter we denote by φ any function from $\mathcal{F}$ such that the compositional inverse $\overline{\varphi}$ exists, $D(\varphi)$ and $D(\overline{\varphi})$ are units, and equation Eq. (2.2) holds for $f = \varphi$.*

With regard to such a φ, we define a mapping $D_\varphi : \mathcal{F} \longrightarrow \mathcal{F}$ by

$$D_\varphi(f) := \frac{D(f)}{D(\varphi)}. \tag{2.5}$$

D_φ is the function-algebraic version of the *Lie derivative* that Todorov used in his paper [99]. $(\mathcal{F}, D_\varphi)$ is a differential ring having the same constants as $(\mathcal{F}, D)$. However, with regard to D_φ we have: (D3) $\Leftrightarrow$ (D4) $\Leftrightarrow$ $D(\varphi) = 1$. Therefore, D_φ satisfies the chain rule if and only if $D_\varphi = D$, that is, in the *trivial* case $\varphi = \iota + c, c \in \mathcal{K}$.

The following simple but useful statement concerns the relation between D_φ^n and D^n (*Pourchet's formula* according to [22, p. 220] and [99, p. 223–224]).

Proposition 2.2. $D_\varphi^n(f) = D^n(f \circ \overline{\varphi}) \circ \varphi$ *for all $n \geq 0$.*

Proof. Verify $D^n(f \circ \overline{\varphi}) = D_\varphi^n(f) \circ \overline{\varphi}$ by induction on n. The case $n = 0$ is clear. For the induction step ($n \to n+1$) use (D3):

$$D^{n+1}(f \circ \overline{\varphi}) = (D(D_\varphi^n(f)) \circ \overline{\varphi})) \cdot D(\overline{\varphi}) =: (*).$$

Applying Eq. (2.2) to $D(\overline{\varphi})$ yields $(*) = \frac{D(D_\varphi^n(f))}{D(\varphi)} \circ \overline{\varphi} = D_\varphi^{n+1}(f) \circ \overline{\varphi}$. $\Diamond$

2.2 Exponential and logarithm

For some purposes it will prove convenient to have in $(\mathcal{F}, +, \cdot, \circ, D)$ besides the three identity elements more functions with special properties, the most important examples being the exponential (exp) and its compositional inverse (log). Of course we expect exp and log to have the familiar properties known from analysis, like $D(\exp) = \exp$, $D(\log) \cdot \iota = 1$, and $D^k(\log) = (-1)^{k-1}(k-1)! \, \iota^{-k}$ for all $k \geq 1$. In such a case we call $\mathcal{F}$ an *extended function algebra*.

Items 3 and 4 from Examples 2.1 are extended function algebras. In $\mathbb{R}[[x]]$ (item 2), however, neither $D(\log)$ nor log have counterparts. In Section 7 (and in Chapter II) we shall therefore deal with $\log \circ (1 + \iota)$ and $\exp -1$.

Notation. In an extended function algebra, we prefer to write id (or x) instead of ι. If no misunderstanding is likely, we will occasionally replace $\exp \circ f$ with e^f and also switch from $f \circ g$ to the usual notation $f(g)$, mainly when g is a constant or the identity function.

3 Expansion of higher-order derivatives

In this section, some basic facts regarding the expansion of D_φ^n (see Eq. (1.5) and Todorov [99, 100]) as well as of D^n will be reformulated and set up for an arbitrary function algebra $\mathcal{F}$ with derivation D. The main idea of the presentation is to make it clear that from the beginning these results are linked by an inversion of functions. We will also examine some basic properties of the multivariable polynomials involved, in particular their differential recurrences and their relationship to the Stirling numbers.

3.1 Expansion formulas for D_φ^n and D^n

Proposition 3.1. *Let f be any function from $\mathcal{F}$ and n, k non-negative integers. Then there are $a_{n,k} \in \mathcal{R}_{n-k+1}(\varphi), 0 \leq k \leq n$, such that*

$$D_\varphi^n(f) = \sum_{k=0}^{n} a_{n,k} \cdot D^k(f).$$

The coefficients $a_{n,k}$ are uniquely determined by the recurrence

$$a_{n+1,k} = \frac{a_{n,k-1} + D(a_{n,k})}{D(\varphi)} \quad (1 \leq k \leq n+1),$$

where $a_{0,0} = 1, a_{i,0} = 0 \ (i > 0)$, and $a_{i,j} = 0 \ (0 \leq i < j)$.

Proof. Recall that $D(\varphi)$ is a unit and $(\mathcal{F}, D_\varphi)$ is a differential ring. So, $D_\varphi^n(f)$ is defined for every $n \geq 0$ and can successively be calculated by applying (D1) and (D2) together with the rule $D_\varphi(g^{-1}) = -g^{-2}D_\varphi(g)$ (g a unit). The proof is then carried out by a simple induction on n, the details of which can be omitted here.   ◇

Remark 3.1. $a_{n,n} = D(\varphi)^{-n}$ for all $n \geq 0$.

One obtains by induction that the denominator in $a_{n,k}$ is $D(\varphi)^{2n-1}$.

Corollary 3.2. *Set* $s_{n,k} := D(\varphi)^{2n-1}a_{n,k}$. *Then* $s_{n,k} \in \mathcal{P}_{n-k+1}(\varphi)$ *for* $(n,k) \neq (0,0)$, *and the following recurrence holds:*

$$s_{n+1,k} = -(2n-1)D^2(\varphi)s_{n,k} + D(\varphi) \cdot (s_{n,k-1} + D(s_{n,k})),$$

where $s_{0,0} = D(\varphi)^{-1}$, $s_{1,1} = 1$, $s_{i,0} = 0$ $(i > 0)$, *and* $s_{i,j} = 0$ $(0 \leq i < j)$.

It is natural to ask whether, conversely, D^n can be expanded into a linear combination of the D_φ^k $(k = 0, 1, \ldots, n)$. The next proposition gives a positive answer.

Proposition 3.3. *Let f be any function and n, k non-negative integers. Then, there are* $b_{n,k} \in \mathcal{P}_{n-k+1}(\varphi)$, $0 \leq k \leq n$, *such that*

$$D^n(f) = \sum_{k=0}^{n} b_{n,k} \cdot D_\varphi^k(f).$$

The coefficients $b_{n,k}$ are uniquely determined by the recurrence

$$b_{n+1,k} = D(\varphi) \cdot b_{n,k-1} + D(b_{n,k}) \quad (1 \leq k \leq n+1),$$

where $b_{0,0} = 1$, $b_{i,0} = 0$ $(i > 0)$, *and* $b_{i,j} = 0$ $(0 \leq i < j)$.

Proof. We apply Proposition 3.1 to the compositional inverse $\overline{\varphi}$, thus obtaining $a'_{n,k} \in \mathcal{R}_{n-k+1}(\overline{\varphi})$ so that $D_{\overline{\varphi}}^n(f) = \sum_{k=0}^{n} a'_{n,k} \cdot D^k(f)$. Since according to Pourchet's formula (Proposition 2.2) the left-hand side is equal to $D^n(f \circ \varphi) \circ \overline{\varphi}$, we get by (F3)

$$D^n(f \circ \varphi) = \sum_{k=0}^{n}(a'_{n,k} \circ \varphi) \cdot (D^k(f) \circ \varphi). \tag{3.1}$$

We now replace f by $f \circ \overline{\varphi}$ in Eq. (3.1). Then again Pourchet's formula, applied to the second factor on the right-hand side of Eq. (3.1), yields

$$D^n(f) = \sum_{k=0}^{n}(a'_{n,k} \circ \varphi) \cdot D_\varphi^k(f).$$

Now set $b_{n,k} := a'_{n,k} \circ \varphi$. We then have $b_{0,0} = a'_{0,0} \circ \varphi = 1 \circ \varphi = 1$, likewise $b_{i,0} = 0$ $(i > 0)$, $b_{i,j} = 0$ $(0 \leq i < j)$, and by Eq. (2.2)

$$b_{n+1,k} = a'_{n+1,k} \circ \varphi = (D(\overline{\varphi})^{-1} \circ \varphi) \cdot ((D(a'_{n,k}) \circ \varphi) + (a'_{n,k-1} \circ \varphi))$$

$$= D(\varphi) \cdot (D(a'_{n,k}) \circ \varphi) + D(\varphi) \cdot b_{n,k-1}$$
$$= D(b_{n,k}) + D(\varphi) \cdot b_{n,k-1} \qquad \text{(by the chain rule (D3)).}$$

Finally, $b_{n,k} \in \mathcal{P}_{n-k+1}(\varphi)$ follows from this recurrence by an inductive argument. $\qquad \Diamond$

Remark 3.2. $b_{n,n} = D(\varphi)^n$ for all $n \geq 0$.

3.2 *Fundamental properties of the coefficients*

Let us now have a closer look at the coefficient functions $a_{n,k}$ and $b_{n,k}$. We start with $b_{n,k}$. As it is a polynomial expression in the derivatives $D(\varphi), \ldots, D^{n-k+1}(\varphi)$, we get $b_{n,k}$ from a suitable polynomial family $B_{n,k}$ by substituting $D^j(\varphi)$ in place of the indeterminates X_j, that is, $B_{n,k}^{\varphi} = b_{n,k}$. In the case of $a_{n,k}$ it is likewise clear by Corollary 3.2 that $s_{n,k}$, too, comes from certain polynomials $S_{n,k}$ satisfying $S_{n,k}^{\varphi} = s_{n,k} = D(\varphi)^{2n-1}a_{n,k}$. We therefore define $A_{n,k} := X_1^{-(2n-1)} S_{n,k}$. Then, of course, $A_{n,k}^{\varphi} = a_{n,k}$ holds. Note that $A_{n,k}$ is a Laurent polynomial, and that is especially also true for $S_{0,0}$ (see Corollary 3.2).

The polynomials $S_{n,k}$ ($A_{n,k}$) and $B_{n,k}$ are closely connected.

Proposition 3.4.

(i) $\qquad B_{n,k}^{\varphi} = D(\varphi)^{2n-1} \cdot (S_{n,k}^{\overline{\varphi}} \circ \varphi), \qquad B_{n,k}^{\varphi} = A_{n,k}^{\overline{\varphi}} \circ \varphi$

(ii) $\qquad S_{n,k}^{\varphi} = D(\varphi)^{2n-1} \cdot (B_{n,k}^{\overline{\varphi}} \circ \varphi), \qquad A_{n,k}^{\varphi} = B_{n,k}^{\overline{\varphi}} \circ \varphi$

Proof. (i): From the proof of Proposition 3.3 we obtain

$$B_{n,k}^{\varphi} = a'_{n,k} \circ \varphi = A_{n,k}^{\overline{\varphi}} \circ \varphi.$$

Now note the generalized inversion rule obtained by induction from Eq. (2.2):

$$D(\overline{\varphi})^m \circ \varphi = D(\varphi)^{-m} \quad \text{for all integers } m \geq 0.$$

Hence $S_{n,k}^{\overline{\varphi}} \circ \varphi = (D(\overline{\varphi})^{2n-1} \circ \varphi) \cdot (A_{n,k}^{\overline{\varphi}} \circ \varphi) = D(\varphi)^{-(2n-1)} \cdot B_{n,k}^{\varphi}$.
(ii): Replace φ in (i) by $\overline{\varphi}$. $\qquad \Diamond$

From the foregoing we gather the following special values:

$$A_{n,n} = X_1^{-n}, \quad S_{n+1,n+1} = B_{n,n} = X_1^n \qquad (n \geq 0),$$

$$A_{i,0} = S_{i,0} = B_{i,0} = 0 \qquad\qquad (i > 0),$$
$$A_{i,j} = S_{i,j} = B_{i,j} = 0 \qquad\qquad (0 \le i < j).$$

What still remains to be done is transforming the differential recurrences for $a_{n,k}$, $s_{n,k}$, $b_{n,k}$ into recurrences for the corresponding polynomials $A_{n,k}$, $S_{n,k}$, $B_{n,k}$. Consider the derivative

$$D(a_{n,k}) = D(A_{n,k}(D(\varphi), \dots, D^{n-k+1}(\varphi))).$$

Applying Eq. (2.3) to the right-hand side, we obtain

$$D(a_{n,k}) = \sum_{j=1}^{n-k+1} \frac{\partial A_{n,k}}{\partial X_j}\left(D(\varphi), \dots, D^{n-k+1}(\varphi)\right) \cdot D^{j+1}(\varphi)$$

$$= \left(\sum_{j=1}^{n-k+1} X_{j+1}\frac{\partial A_{n,k}}{\partial X_j}\right)^{\varphi}.$$

$D(s_{n,k})$ and $D(b_{n,k})$ resolve in the same manner. So, according to Proposition 3.1, Corollary 3.2, and Proposition 3.3 we have the following

Proposition 3.5.

(i) $\quad A_{n+1,k} = \dfrac{1}{X_1}\left(A_{n,k-1} + \displaystyle\sum_{j=1}^{n-k+1} X_{j+1}\frac{\partial A_{n,k}}{\partial X_j}\right),$

(ii) $\quad S_{n+1,k} = -(2n-1)X_2 S_{n,k} + X_1\left(S_{n,k-1} + \displaystyle\sum_{j=1}^{n-k+1} X_{j+1}\frac{\partial S_{n,k}}{\partial X_j}\right),$

(iii) $\quad B_{n+1,k} = X_1 B_{n,k-1} + \displaystyle\sum_{j=1}^{n-k+1} X_{j+1}\frac{\partial B_{n,k}}{\partial X_j}.$

It follows (by induction) from Proposition 3.5 that these polynomials have integral coefficients. We denote by $\sigma_{n,k}(r_1, \dots, r_{n-k+1})$ and $\beta_{n,k}(r_1, \dots, r_{n-k+1})$ the coefficients of $X_1^{r_1} \cdots X_{n-k+1}^{r_{n-k+1}}$ in $S_{n,k}$ and in $B_{n,k}$, respectively, thus obtaining

Corollary 3.6.

(i) $\quad S_{n,k} = \displaystyle\sum_{\mathbb{P}(2n-1-k, n-1)} \sigma_{n,k}(r_1, \dots, r_{n-k+1}) X_1^{r_1} \cdots X_{n-k+1}^{r_{n-k+1}},$

(ii) $B_{n,k} = \sum\limits_{\mathbb{P}(n,k)} \beta_{n,k}(r_1, \ldots, r_{n-k+1}) X_1^{r_1} \cdots X_{n-k+1}^{r_{n-k+1}}.$

Proof. (i): By induction on n. For $n = 1$ we have the degenerate case of a $(0,0)$-partition type, $r_1 = 0$. Thus $S_{1,1} = 1$ can be achieved by choosing $\sigma_{1,1}(0) = 1$. The induction step $(n \to n+1)$ is carried out by examining the partition types produced by the terms $X_2 S_{n,k}$, $X_1 S_{n,k-1}$, and $X_1 X_{j+1} \frac{\partial S_{n,k}}{\partial X_j}$ in part (ii) of Proposition 3.5. Each of them makes $\sum r_i \, (= n-1)$ increase by 1 and makes $\sum i r_i \, (= 2n-1-k)$ increase by 2, which gives the appropriate $(2n+1-k, n)$-partition types for $S_{n+1,k}$. — (ii): Obviously by a similar argument.   ◇

Remark 3.3. We already know that $S_{n,n} = X_1^{n-1}$. Taking $k = n$ in Proposition 3.5 (ii) then yields $S_{n,n-1} = -\binom{n}{2} X_1^{n-2} X_2$ $(n \geq 2)$. Todorov [100] has also calculated $S_{n,n-2}$ and $S_{n,n-3}$ this way.

Remark 3.4. As a consequence of Corollary 3.6, $S_{n,k}$ is homogeneous of degree $n-1$ and isobaric of degree $2n-1-k$, while $B_{n,k}$ is homogeneous of degree k and isobaric of degree n.

We now define integers $s_1(n,k) := A_{n,k}(1, \ldots, 1) = S_{n,k}(1, \ldots, 1)$ and $s_2(n,k) := B_{n,k}(1, \ldots, 1)$, that is, $s_1(n,k)$, $s_2(n,k)$ are the sums of the coefficients of $A_{n,k}$ $(S_{n,k})$ and $B_{n,k}$, respectively.

Proposition 3.7. *Let n, k be integers, $0 \leq k \leq n$. Then, $s_1(n,k)$ are the signed Stirling numbers of the first kind, and $s_2(n,k)$ are the Stirling numbers of the second kind:*

(i) $s_1(n,k) = (-1)^{n-k} \begin{bmatrix} n \\ k \end{bmatrix}$,   (ii) $s_2(n,k) = \begin{Bmatrix} n \\ k \end{Bmatrix}.$

Proof. (i): From the special values above we gather $s_1(0,0) = 1$, $s_1(i,0) = 0$ $(i > 0)$, and $s_1(i,j) = 0$ $(0 \leq i < j)$. It suffices to show that s_1 satisfies the recurrence $s_1(n+1,k) = s_1(n, k-1) - n s_1(n,k)$, which defines the Stirling numbers of the first kind (see e.g. [77, p. 33]). We use Corollary 3.6 (i). Consider first

$$X_{j+1} \frac{\partial S_{n,k}}{\partial X_j} = \sum\limits_{\mathbb{P}(2n-1-k, n-1)} r_j \, \sigma_{n,k}(r_1, \ldots, r_{n-k+1}) \times$$
$$\times X_1^{r_1} \cdots X_j^{r_j - 1} X_{j+1}^{r_{j+1}+1} \cdots X_{n-k+1}^{r_{n-k+1}}.$$

Replacing all indeterminates by 1 and then taking the sum from $j = 1$ to $n - k + 1$ yields

$$\sum_{j=1}^{n-k+1} \sum_{\mathbb{P}(2n-1-k,n-1)} r_j \, \sigma_{n,k}(r_1, \ldots, r_{n-k+1}) = s_1(n, k) \sum_{j=1}^{n-k+1} r_j.$$

Observing $r_1 + \cdots + r_{n-k+1} = n - 1$ we get by Proposition 3.5 (ii)

$$
\begin{aligned}
s_1(n + 1, k) &= -(2n - 1)s_1(n, k) + s_1(n, k - 1) + (n - 1)s_1(n, k) \\
&= s_1(n, k - 1) - ns_1(n, k).
\end{aligned}
$$

(ii): The recurrence $s_2(n + 1, k) = s_2(n, k - 1) + ks_2(n, k)$ that defines the Stirling numbers of the second kind, can be verified by a similar argument using (iii) from Proposition 3.5. $\diamond$

Examples 3.1. (i) We consider some special cases in an extended function algebra:

(1) $\quad A_{n,k}^{\exp} = s_1(n, k) \cdot \exp^{-n}$ (1') $\quad B_{n,k}^{\log} = s_1(n, k) \cdot \mathrm{id}^{-n}$

(2) $\quad B_{n,k}^{\exp} = s_2(n, k) \cdot \exp^{k}$ (2') $\quad A_{n,k}^{\log} = s_2(n, k) \cdot \mathrm{id}^{k}$.

(1') and (2') immediately follow by Proposition 3.4 from (1) and (2), respectively. It is enough to perform the calculation for (1):

$$
\begin{aligned}
A_{n,k}^{\exp} &= A_{n,k}(D(\exp), \ldots, D^{n-k+1}(\exp)) \\
&= D(\exp)^{-(2n-1)} \cdot S_{n,k}(\exp, \ldots, \exp) \\
&= \exp^{-(2n-1)} \cdot \exp^{n-1} \cdot S_{n,k}(1, \ldots, 1) \quad \text{(Remark 3.4)} \\
&= s_1(n, k) \cdot \exp^{-n}. \quad \text{(Definition } s_1, \text{ Proposition 3.7)}
\end{aligned}
$$

(ii) Since Proposition 3.7 tells us that $s_1(n, k)$ are in fact the signed Stirling numbers of the first kind, we choose $\varphi = \log$ in the expansion of Proposition 3.3, which implies equation Eq. (1.2) in the form

$$D^n(f) = \sum_{k=1}^{n} B_{n,k}^{\log} \cdot D_{\log}^{k}(f) = \mathrm{id}^{-n} \sum_{k=1}^{n} s_1(n, k)(\mathrm{id} \cdot D)^{k}(f).$$

Analogously combining (2) and (2') with Proposition 3.1, the reader should verify also the following expansion formula that corresponds to Eq. (1.1):

$$(\mathrm{id} \cdot D)^n(f) = D^n_{\log}(f) = \sum_{k=1}^n s_2(n,k) \cdot \mathrm{id}^k \cdot D^k(f).$$

(iii) It may be of interest to show how some special Stirling numbers can be directly calculated within the machinery of an extended function algebra. Let us, for instance, compute $s_1(n,1)$. By (1) and Proposition 3.4

$$s_1(n,1) = \exp^n \cdot A^{\exp}_{n,1} = \exp^n \cdot (B^{\log}_{n,1} \circ \exp) = \exp^n \cdot (D^n(\log) \circ \exp).$$

Observing that $s_1(n,k) \in \mathcal{K}$, we obtain

$$\begin{aligned} s_1(n,1) &= s_1(n,1) \circ 0 \\ &= 1 \cdot (D^n(\log) \circ 1) = (-1)^{n-1}(n-1)!. \end{aligned}$$

I use the term *multivariate Stirling polynomial* (MSP) to denote both $S_{n,k}$ (MSP of the first kind) and $B_{n,k}$ (MSP of the second kind). Proposition 3.7 may be regarded as a good reason for this eponymy (see also my comments in Section 1 concerning notation and terminology).

Remark 3.5. The $B_{n,k}$ are widely known as *partial* exponential Bell polynomials (see e. g. [18, 22], also the explicit formula (1.4)). Their *complete* form is defined by $B_n := \sum_{k=1}^n B_{n,k}$. Applying Proposition 3.5 (iii) to each term of this sum gives the differential recurrence $B_{n+1} = X_1 B_n + \sum_{j=1}^n X_{j+1} \frac{\partial B_n}{\partial X_j}$, which originally has been studied by Bell [6]. In [77, p. 49] the complete Bell polynomials are tabulated up to $n = 8$.

Remark 3.6. In [51, p. 52 and pp. 481–483] Knuth analyzes, from a combinatorial point of view, the coefficients of $B_{n,k}$ in connection with Eq. (3.1) thus establishing a recurrence for $\beta_{n,k}$. Set $\beta_{n,k}(\ldots) = 0$ for partition types $\notin \mathbb{P}(n,k)$. Then $\beta_{1,1}(1) = 1$, and for every $(r_1, \ldots, r_{n-k+2}) \in \mathbb{P}(n+1,k)$:

$$\beta_{n+1,k}(r_1, \ldots, r_{n-k+2}) = \beta_{n,k-1}(r_1 - 1, r_2, \ldots, r_{n-k+2}) +$$
$$\sum_{j=1}^{n-k+1} (r_j + 1)\beta_{n,k}(\ldots, r_j + 1, r_{j+1} - 1, \ldots).$$

This could also be obtained more formally by combining Proposition 3.5 (iii) with Corollary 3.6 (ii).

I give here without proof also a recurrence for $\sigma_{n,k}$ (to be obtained by using Proposition 3.5 (ii) and Corollary 3.6 (i)). If we agree in an analogous way to let $\sigma_{n,k}$ vanish for partition types $\notin \mathbb{P}(2n-1-k, n-1)$, then we have $\sigma_{1,1}(1) = 1$, and for every $(r_1, \ldots, r_{n-k+2}) \in \mathbb{P}(2n+1-k, n)$:

$$
\begin{aligned}
\sigma_{n+1,k}(r_1, \ldots, r_{n-k+2}) &= \sigma_{n,k-1}(r_1 - 1, r_2, \ldots, r_{n-k+2}) \\
&+ (r_1 - 2n + 1)\sigma_{n,k}(r_1, r_2 - 1, \ldots, r_{n-k+1}) \\
&+ \sum_{j=2}^{n-k+1} (r_j + 1)\sigma_{n,k}(r_1 - 1, \ldots, r_j + 1, r_{j+1} - 1, \ldots, r_{n-k+1}).
\end{aligned}
$$

4 A brief summary on Bell polynomials

Replacing the coefficient $a'_{n,k} \circ \varphi$ in Eq. (3.1) (cf. the proof of Proposition 3.3) by $B^{\varphi}_{n,k}$, we obtain for $n \geq 0$

$$
D^n(f \circ \varphi) = \sum_{k=0}^{n} B^{\varphi}_{n,k} \cdot (D^k(f) \circ \varphi). \tag{4.1}
$$

This is, in function-algebraic notation, the well-known *Faà di Bruno formula* (1.3) for the higher derivatives of a composite function. Though it has been known for a long time, it may, from a systematic point of view, appear appropriate to briefly examine here some of the related classical results on $B_{n,k}$ within our framework.

Let $F \in \mathcal{F}[X]$. We denote by $[F(\varphi) \mid \varphi = 0]$ the result of substituting 0 for φ in the monomial products φ^j of $F(\varphi)$ with $j \geq 1$. Example: Let $F = \varphi + X^2 + 3(1 - X)D(\varphi)$; then $[F(\varphi) \mid \varphi = 0] = 3D(\varphi)$.

Proposition 4.1. *For $1 \leq k \leq n$ we have*

$$
B^{\varphi}_{n,k} = \frac{1}{k!} \left[D^n(\varphi^k) \mid \varphi = 0 \right].
$$

Proof. By Eq. (4.1) we obtain

$$
\begin{aligned}
D^n(\varphi^k) = D^n(\iota^k \circ \varphi) &= \sum_{j=0}^{n} B^{\varphi}_{n,j} \cdot (D^j(\iota^k) \circ \varphi) \\
&= \sum_{j=1}^{k} B^{\varphi}_{n,j} \cdot (k)_j \, \varphi^{k-j}
\end{aligned}
$$

$$= k! \cdot B_{n,k}^{\varphi} + \varphi \sum_{j=1}^{k-1} B_{n,j}^{\varphi} \cdot (k)_j \, \varphi^{k-1-j}. \tag{4.2}$$

Taking φ to 0 gives the asserted. $\diamond$

Proposition 4.2. *For* $1 \leq k \leq n$ *we have*

$$B_{n,k}^{\varphi} = \frac{1}{k!} \sum_{j=1}^{k} (-1)^{k-j} \binom{k}{j} \varphi^{k-j} D^n(\varphi^j).$$

Proof. We rewrite Eq. (4.2) in the form

$$D^n(\varphi^k) = \sum_{j=1}^{k} \binom{k}{j} j! B_{n,j}^{\varphi} \, \varphi^{k-j}.$$

Applying binomial inversion (cf. [3, p. 96-97]) to the equivalent equation

$$\varphi^{-k} D^n(\varphi^k) = \sum_{j=1}^{k} \binom{k}{j} \varphi^{-j} j! B_{n,j}^{\varphi}$$

yields

$$\varphi^{-k} k! B_{n,k}^{\varphi} = \sum_{j=1}^{k} (-1)^{k-j} \binom{k}{j} \varphi^{-j} D^n(\varphi^j).$$

Division by $\varphi^{-k} k!$ then gives the asserted. $\diamond$

Now recall the definition of the subset function β in Eq. (1.12).

Proposition 4.3. *We have* $\beta_{n,k}(r_1, \ldots, r_{n-k+1}) = \beta(r_1, \ldots, r_{n-k+1})$ *for all* $(r_1, \ldots, r_{n-k+1}) \in \mathbb{P}(n, k)$, *that is,*

$$B_{n,k} = \sum_{\mathbb{P}(n,k)} \frac{n!}{r_1! \cdots r_{n-k+1}! \cdot 1!^{r_1} \cdots (n-k+1)!^{r_{n-k+1}}} X_1^{r_1} \cdots X_{n-k+1}^{r_{n-k+1}}.$$

Proof. It follows from the Leibniz rule (2.4)

$$D^n(\varphi^k) = \sum_{\substack{j_1+\cdots+j_k=n \\ j_1,\ldots,j_k \geq 0}} \frac{n!}{j_1!\cdots j_k!} \, D^{j_1}(\varphi) \cdots D^{j_k}(\varphi).$$

From this we get by Proposition 4.1

$$B_{n,k} = \frac{1}{k!} \sum_{\substack{j_1+\cdots+j_k=n \\ j_1,\ldots,j_k \geq 1}} \frac{n!}{j_1!\cdots j_k!} \, X_{j_1} \cdots X_{j_k}. \tag{4.3}$$

Denote by r_m the number of j's equal to $m \in \{1,\ldots,n-k+1\}$. Then, each sequence $(j_1,\ldots,j_k)$ in (4.3) is obtained from its corresponding linearly ordered k-tuple $i_1 \leq \cdots \leq i_k$ by $\frac{k!}{r_1!\cdots r_{n-k+1}!}$ permutations. Hence (4.3) becomes

$$B_{n,k} = \sum_{\substack{i_1+\cdots+i_k=n \\ 1\leq i_1\leq\ldots\leq i_k}} \frac{1}{r_1!\cdots r_{n-k+1}!} \cdot \frac{n!}{i_1!\cdots i_k!} \, X_{i_1} \cdots X_{i_k},$$

where $i_1!i_2!\cdots i_k! = 1!^{r_1} \cdot 2!^{r_2} \cdots (n+k+1)!^{r_{n-k+1}}$. This yields the asserted equation, and $\beta_{n,k}$ (the coefficient function of $B_{n,k}$ according to Corollary 3.6) is shown to agree with β on $\mathbb{P}(n,k)$. $\Diamond$

Remark 4.1. Some historical comments related to Faà di Bruno's formula are given in [42]. One example that 'deserves to be better known' (Johnson), is a formula stated by G. Scott (1861) (cf. [91] and [42, p. 220]). Proposition 4.1 reformulates it in function-algebraic terms. According to [99], the expression for $B_{n,k}^\varphi$ given in Proposition 4.2 is due to J. Bertrand [9, p. 140]). Instead of 'a not so easy induction' (Todorov), its verification needs merely applying binomial inversion to Scott's formula. Finally, taking $\varphi = \exp$ makes Bertrand's formula into a well-known explicit expression for the Stirling numbers of the second kind (cf. [3, p. 97]): $s_2(n,k) = \frac{1}{k!} \sum_{j=1}^{k} (-1)^{k-j} \binom{k}{j} j^n$.

Three corollaries will be useful in later sections.

Because of the relatively simple structure of the $B_{n,k}$ the partial derivatives in Proposition 3.5 (iii) can be given a closed non-differential form.

Corollary 4.4.

$$\frac{\partial B_{n,k}}{\partial X_j} = \binom{n}{j} B_{n-j,k-1} \qquad (1 \le j \le n - k + 1).$$

Proof. The assertion follows by applying $\frac{\partial}{\partial X_j}$ to the explicit expression of $B_{n,k}$ in Proposition 4.3. Observe that $\frac{\partial}{\partial X_j}$ takes each (n,k)-partition type into a $(n-j, k-1)$-partition type. The details are left to the reader. $\diamond$

Corollary 4.5.

$$B_{n,k} = \sum_{r=0}^{k} \binom{n}{r} X_1^r B_{n-r,k-r}(0, X_2, \ldots, X_{n-k+1}).$$

Proof. Immediate from Proposition 4.3. See also Comtet [22, p. 136]. $\diamond$

Notation. (i) The right-hand side of the equation in Corollary 4.5 gives rise to the notation $\widetilde{B}_{n,k} := B_{n,k}(0, X_2, \ldots, X_{n-k+1})$. We call $\widetilde{B}_{n,k}$ *associated Bell polynomial* (or, *associated* MSP of the second kind). The coefficients of $\widetilde{B}_{n,k}$ count only partitions with no singleton blocks. Note that $\tilde{s}_2(n, k) := \widetilde{B}_{n,k}(1, \ldots, 1)$ are the well-known associated Stirling numbers of the second kind [36, 77].

(ii) We call *(unsigned) Lah polynomial* the expression

$$L_{n,k}^+ := P_\omega = \sum_{\mathbb{P}(n,k)} \omega(r_1, \ldots, r_{n-k+1}) X_1^{r_1} \ldots X_{n-k+1}^{r_{n-k+1}},$$

where ω is the order function in Eq. (1.10). We have

$$L_{n,k}^+(1, \ldots, 1) = \frac{n!}{k!} \binom{n-1}{k-1} =: l^+(n, k).$$

Let $l(n, k)$ denote the signed Lah numbers $(-1)^n l^+(n, k)$. Then

$$L_{n,k}^+((-1)^1, (-1)^2, \ldots, (-1)^{n-k+1}) = l(n, k),$$

which follows from the observation that $r_1 + r_3 + r_5 + \ldots \equiv n \pmod 2$ holds whenever $(r_1, \ldots, r_{n-k+1}) \in \mathbb{P}(n, k)$.

Corollary 4.6. $L_{n,k}^+ = B_{n,k}(1! X_1, 2! X_2, \ldots, (n-k+1)! X_{n-k+1}).$

Proof. Immediate from Proposition 4.3. See also Comtet [22, p. 134]. $\diamond$

5 Inversion formulas and recurrences

We now establish some statements concerning inversion as well as recurrence relations of $A_{n,k}$ and $B_{n,k}$. The first one is the polynomial analogue of the well-known inversion law satisfied by the Stirling numbers of the first and second kind (see Eq. (1.9)).

Theorem 5.1 (Inversion Law). *For all $n \geq k \geq 1$*

$$\sum_{j=k}^{n} A_{n,j} B_{j,k} = \delta_{nk} \qquad and \qquad \sum_{j=k}^{n} B_{n,j} A_{j,k} = \delta_{nk}.$$

Remark 5.1. Defining lower triangular matrices $\mathfrak{A}_n := (A_{i,j})_{1 \leq i,j \leq n}$ and $\mathfrak{B}_n := (B_{i,j})_{1 \leq i,j \leq n}$ we can rewrite the statements of the theorem as matrix inversion formulas, for instance, the first one: $\mathfrak{A}_n \mathfrak{B}_n = \mathfrak{I}_n$ (identity matrix) for every $n \geq 1$ (which may also be equivalently expressed by means of differential terms: $\mathfrak{A}_n^\varphi \mathfrak{B}_n^\varphi = \mathfrak{I}_n$). In the special case $X_1 = X_2 = \ldots = 1$, where the Stirling numbers of the first and second kind are the entries of $\mathfrak{A}_n$ and $\mathfrak{B}_n$, respectively, both matrices can be considered as transformation matrices connecting the linearly independent polynomial sequences $(x^1, \ldots, x^n)$ and $((x)_1, \ldots, (x)_n)$ (cf. [3]). Unfortunately, there is much to suggest that this method does not work in our general case. The following proof therefore makes no use of it.

Proof. We prove the first equation of Theorem 5.1. Suppose $1 \leq k \leq n$ and denote by $d_{n,k}$ the sum $\sum_{j=k}^{n} a_{n,j} b_{j,k}$. We need to show that $d_{n,k} = \delta_{nk}$. This is clear for $n = 1$. Now we proceed by induction on n using the differential recurrences in the Propositions 3.1 and 3.3. First, observe that applying D to both sides of the induction hypothesis yields $D(d_{n,k}) = D(\delta_{nk}) = 0$, whence

$$\sum_{j=k}^{n} D(a_{n,j}) b_{j,k} = -\sum_{j=k}^{n} a_{n,j} D(b_{j,k}). \tag{5.1}$$

We then have

$$d_{n+1,k} = a_{n+1,n+1} \, b_{n+1,k} + \sum_{j=k}^{n} a_{n+1,j} \, b_{j,k}$$

$$= \frac{a_{n,n}}{D(\varphi)}\, b_{n+1,k} + \frac{1}{D(\varphi)} \sum_{j=k}^{n}(a_{n,j-1}\, b_{j,k} + D(a_{n,j})b_{j,k})$$

$$= a_{n,n}\left(b_{n,k-1} + \frac{D(b_{n,k})}{D(\varphi)}\right) + \frac{1}{D(\varphi)} \sum_{j=k}^{n}(a_{n,j-1}D(\varphi)b_{j-1,k-1} +$$

$$+ a_{n,j-1}D(b_{j-1,k})) + \frac{1}{D(\varphi)} \sum_{j=k}^{n} D(a_{n,j})b_{j,k}.$$

Replacing the last sum by the right-hand side of Eq. (5.1), we obtain after a short computation:

$$d_{n+1,k} = a_{n,n}\, b_{n,k-1} + a_{n,n}\frac{D(b_{n,k})}{D(\varphi)} + \sum_{j=k}^{n} a_{n,j-1}\, b_{j-1,k-1} - a_{n,n}\frac{D(b_{n,k})}{D(\varphi)}$$

$$= \sum_{j=k-1}^{n} a_{n,j}\, b_{j,k-1} = \delta_{n(k-1)} = \delta_{(n+1)k}. \qquad \diamond$$

We conclude from Theorem 5.1 a statement that generalizes 'Stirling inversion' for sequences of real numbers; cf. [3, Corollary 3.38 (ii)].

Corollary 5.2 (Inversion of sequences). *Let $\mathcal{E}$ be an arbitrary overring of $\mathbb{Z}[X_1,\ldots,X_{n-k+1}]$, $P_0, P_1, P_2, \ldots$ and $Q_0, Q_1, Q_2, \ldots$ any sequences in $\mathcal{E}$. Then the following conditions are equivalent:*

(i) $\qquad P_n = \sum_{k=0}^{n} B_{n,k}Q_k \quad$ *for all $n \geq 0$,*

(ii) $\qquad Q_n = \sum_{k=0}^{n} A_{n,k}P_k \quad$ *for all $n \geq 0$.*

Examples 5.1. (i) Theorem 5.1 implies the above mentioned special inversion law for the Stirling numbers $s_1(n,k) = A_{n,k}(1,\ldots,1)$ and $s_2(n,k) = B_{n,k}(1,\ldots,1)$ of the first and second kind (see Proposition 3.7):

$$\sum_{j=k}^{n} s_1(n,j)s_2(j,k) = \delta_{nk} \qquad (1 \leq k \leq n).$$

(ii) The signed Lah numbers are known to be self-inverse:

$$\sum_{j=k}^{n} l(n,j)l(j,k) = \delta_{nk} \qquad (1 \le k \le n).$$

In order to prove this, set $\varphi(x) = -\frac{x}{1+x}$. Since φ is involutory ($\overline{\varphi} = \varphi$) and $\varphi(0) = 0$, Proposition 3.4 (ii) yields

$$A^{\varphi}_{n,k}(0) = (B^{\overline{\varphi}}_{n,k} \circ \varphi)(0) = B^{\varphi}_{n,k} \circ \varphi(0) = B^{\varphi}_{n,k}(0).$$

It is easily seen that $D^j(\varphi)(0) = (-1)^j j!$ for all $j \ge 1$. Thus by Corollary 4.6 we have $B^{\varphi}_{n,k}(0) = B_{n,k}(-1!, 2!, -3!, 4!, \ldots) = l(n,k)$. Applying Theorem 5.1 then gives the desired result.

Theorem 5.3. *For* $1 \le k \le n$

$$\text{(i)} \qquad A_{n,k} = B_{n,k}(A_{1,1}, \ldots, A_{n-k+1,1}),$$

$$\text{(ii)} \qquad B_{n,k} = A_{n,k}(A_{1,1}, \ldots, A_{n-k+1,1}).$$

Proof. (i): By Proposition 3.4 (ii) $A^{\varphi}_{n,k} = B^{\overline{\varphi}}_{n,k} \circ \varphi$. Corollary 3.6 yields

$$B^{\overline{\varphi}}_{n,k} \circ \varphi = \sum_{\mathbb{P}(n,k)} \beta_{n,k}(r_1, \ldots, r_{n-k+1}) \cdot \prod_{j=1}^{n-k+1} (D^j(\overline{\varphi}) \circ \varphi)^{r_j}.$$

Setting $k = 1$ we get $D^j(\overline{\varphi}) \circ \varphi = B^{\overline{\varphi}}_{j,1} \circ \varphi = A^{\varphi}_{j,1}$ for every $j \ge 1$, hence

$$A^{\varphi}_{n,k} = \sum_{\mathbb{P}(n,k)} \beta_{n,k}(r_1, \ldots, r_{n-k+1})(A^{\varphi}_{1,1})^{r_1} \cdots (A^{\varphi}_{n-k+1,1})^{r_{n-k+1}}$$

$$= B_{n,k}(A^{\varphi}_{1,1}, \ldots, A^{\varphi}_{n-k+1,1})$$

$$= B_{n,k}(A_{1,1}, \ldots, A_{n-k+1,1})^{\varphi}.$$

(ii): Similarly by Proposition 3.4 (i). $\diamond$

The equations of Theorem 5.3 can be rewritten as statements about the polynomials $S_{n,k}$.

Corollary 5.4.

$$\text{(i)} \qquad S_{n,k} = X_1^{k-1} \cdot B_{n,k}(S_{1,1}, \ldots, S_{n-k+1,1}),$$

$$\text{(ii)} \qquad B_{n,k} = X_1^{2k-n} \cdot S_{n,k}(S_{1,1}, \ldots, S_{n-k+1,1}).$$

Proof. We show only (ii). It follows from Theorem 5.3 (ii)

$$B_{n,k} = \frac{1}{(A_{1,1})^{2n-1}} \, S_{n,k}\left(\frac{S_{1,1}}{X_1^1}, \dots, \frac{S_{n-k+1,1}}{X_1^{2(n-k+1)-1}}\right) = X_1^{2n-1} \times$$

$$\sum_{\mathbb{P}(2n-1-k,n-1)} \left(\sigma_{n,k}(r_1, r_2, \dots) \cdot \prod_{j=1}^{n-k+1} (S_{j,1})^{r_j} \cdot X_1^{-\sum_{j=1}^{n-k+1}(2j-1)r_j}\right)$$

$$= X_1^{2n-1} \cdot X_1^{-(3n-2k-1)} \cdot S_{n,k}(S_{1,1}, \dots, S_{n-k+1,1}).$$

In the last two lines Corollary 3.6 (ii) has been used.   ◇

Remark 5.2. Though equations (i) from both Theorem 5.3 and Corollary 5.4 look like recurrences, their practical (computational) value is rather poor, insofar as they work recursively only for $2 \leq k \leq n$. For example, one can actually get $S_{5,3}$ by evaluating $X_1^2 \cdot B_{5,3}(S_{1,1}, S_{2,1}, S_{3,1})$ to $45 X_1^2 X_2^2 - 10 X_1^3 X_3$. It should be noted, however, that the very first members of each generation, $S_{n,1}$, are the most complicated, and in this case, of all things, (i) yields the empty statement $S_{n,1} = B_{n,1}(S_{1,1}, \dots, S_{n,1})$, where $B_{n,1} = X_n$.

Remark 5.3. Through Corollary 5.4, the particular role of X_1 becomes evident. The exponents appearing here have a combinatorial meaning. Given any $(r_1, \dots, r_{n-k+1}) \in \mathbb{P}(2n-1-k, n-1)$, one has $r_1 \geq k-1 \geq 0$. For partition types from $\mathbb{P}(n,k)$, a corresponding, however possibly negative lower bound holds: $r_1 \geq 2k - n$.

Example 5.2. It has already been illustrated that putting $X_j = 1$ ($j = 1, 2, \dots$) converts MSP relations into statements about Stirling numbers. So we may ask what in this regard Theorem 5.3 (i) is about. By Proposition 3.7 we obtain a neat identity for the signed Stirling numbers of the first kind:

$$s_1(n,k) = B_{n,k}(s_1(1,1), \dots, s_1(n-k+1,1)).   (5.2)$$

We know that $B_{n,k} = P_\beta$ (Proposition 4.3) and $s_1(j,1) = (-1)^{j-1}(j-1)!$ (Examples 3.1 (iii)). Hence a straightforward evaluation of the right-hand side of (5.2) eventually yields $s_1(n,k) = (-1)^{n-k} c(n,k)$ with $c(n,k) = \sum_{\mathbb{P}(n,k)} \zeta(r_1, \dots, r_{n-k+1})$ (Cauchy's famous enumeration of n-permutations with exactly k cycles by means of the cycle function ζ, (1.11)). Compare equation [3i] in [22, p. 135] for a signless version of Eq. (5.2).

We will now establish *one* recurrence relation that is satisfied by both $A_{n,k}$ and $B_{n,k}$. However, as with Theorem 5.3, the recurrence does not work for $k = 1$, since $A_{j,1}$ and $B_{j,1}$ ($1 \leq j \leq n - k + 1$) are needed as initial values.

Proposition 5.5. *Let* n, k *be integers with* $1 \leq k \leq n$. *Then we have*

$$\text{(i)} \qquad A_{n,k} = \sum_{j=1}^{n-k+1} \binom{n-1}{j-1} A_{j,1} A_{n-j,k-1},$$

$$\text{(ii)} \qquad B_{n,k} = \sum_{j=1}^{n-k+1} \binom{n-1}{j-1} B_{j,1} B_{n-j,k-1}.$$

Proof. (i) can be easily inferred from (ii): We transform B into $B^{\overline{\varphi}}$ and apply $\circ \varphi$ (from the right) on both sides of the equation. Then, Proposition 3.4 (ii) yields the desired statement. $-$ (ii): Eliminating the partial derivative in Proposition 3.5 (iii) by Corollary 4.4 leads to

$$B_{n,k} = X_1 B_{n-1,k-1} + \sum_{j=1}^{n-1-k+1} X_{j+1} \binom{n-1}{j} B_{n-1-j,k-1}$$

$$= \sum_{j=1}^{n-k+1} X_j \binom{n-1}{j-1} B_{n-j,k-1}.$$

The observation $X_j = B_{j,1}$ completes the proof. $\diamond$

Corollary 5.6.

$$\widetilde{B}_{n,k} = \sum_{j=2}^{n-k+1} \binom{n-1}{j-1} X_j \widetilde{B}_{n-j,k-1} \qquad (1 \leq k \leq n).$$

Remark 5.4. Proposition 5.5 (ii) is stated in Charalambides [18] (together with a generating function proof; see ibid., p. 415). In the special case $k = 2$ we have $B_{j,1} B_{n-j,k-1} = X_j X_{n-j}$. Thus (ii) becomes a simple explicit formula for $B_{n,2}$; see also [77, p. 48].

Remark 5.5. From Corollary 5.6 it follows by an inductive argument that $\widetilde{B}_{2n-l,n} = 0$, if $l \geq 1$, and $\widetilde{B}_{2n,n} = (1 \cdot 3 \cdot 5 \cdots (2n-1)) X_2^n$.

Examples 5.3 (Proposition 5.5). The substitution $X_j = 1$ makes (i) and (ii) into statements about Stirling numbers. Observe $s_2(j, 1) = 1$ and $s_1(j, 1) = (-1)^{j-1}(j-1)!$ (Examples 3.1 (iii)). Then, a little calculation yields the following well-known identities (cf. [77, p. 42-43] and [51, p. 68]):

$$\begin{bmatrix} n+1 \\ k+1 \end{bmatrix} = \sum_{j=k}^{n} \frac{n!}{j!} \begin{bmatrix} j \\ k \end{bmatrix}, \qquad \begin{Bmatrix} n+1 \\ k+1 \end{Bmatrix} = \sum_{j=k}^{n} \binom{n}{j} \begin{Bmatrix} j \\ k \end{Bmatrix}.$$

6 Explicit formulas for $S_{n,k}$

We now pass to the task of finding a fully explicit expression for $S_{n,k}$. At first glance it seems a viable idea to get $S_{n,k}$ by elimination from the inversion law (Theorem 5.1), which is a linear system already in triangular form. In fact, one can verify by induction that every MSP of the first kind can be expressed in terms of Bell polynomials. For instance, in the leading case $k = 1 < n$ we obtain

$$S_{n,1} = -X_1^{n-2} B_{n,1} +$$
$$\sum_{r=1}^{n-2} (-1)^{r+1} \sum_{1 < j_1 < \cdots < j_r < n} X_1^{(n-2)-(j_1+\cdots+j_r)} B_{n,j_r} B_{j_r,j_{r-1}} \cdots B_{j_1,1}. \tag{6.1}$$

A quite similar version of Eq. (6.1) (with $X_1 = 1$) together with an evaluation of the case $n = 5$ is mentioned by Figueroa and Gracia-Bondía in connection with the antipode on a Hopf algebra (cf. [28, equation (7.8)]).

While it seems dubious if and how Eq. (6.1) could be further simplified so as to become practicable, it clearly underlines at least again the particular role of X_1 already observed in Remark 5.3. So, it may appear a promising idea trying to expand $S_{n,k}$ into a finite series

$$C_{n,k,1} X_1 + C_{n,k,2} X_1^2 + C_{n,k,3} X_1^3 + \cdots,$$

whose coefficients $C_{n,k,r}$ do neither contain X_1 nor products of two or more Bell polynomials.

The main result we are going to establish in Theorem 6.1 may be regarded as a non-trivial counterpart of Corollary 4.5; it indeed expresses *all* $S_{n,k}$ (and consequently $A_{n,k}$) in terms of associated Bell polynomials.

Theorem 6.1. *For $n \geq k \geq 1$*

$$S_{n,k} = \sum_{r=k-1}^{n-1} (-1)^{n-1-r} \binom{2n-2-r}{k-1} X_1^r \widetilde{B}_{2n-1-k-r,n-1-r}.$$

Proof. The proof is divided into two parts. First we will show by induction[2] that there are polynomials $C_{n,k,r} \in \mathbb{Z}[X_2, \ldots, X_{n-k+1}]$ such that

$$S_{n,k} = \sum_{r=k-1}^{n-1} C_{n,k,r} X_1^r, \tag{6.2}$$

where the $C_{n,k,r}$ are uniquely determined by a certain differential recurrence. Therefore, in the second step, it remains to show that the recurrence is satisfied by the coefficients of X_1^r in the asserted equation of the theorem.

1. For $k = n$ we have $S_{n,n} = X_1^{n-1}$ (Remark 3.3), and taking $C_{n,n,n-1} = 1$ satisfies Eq. (6.2). This includes the case $n = 1$. Now let $n \geq 1$ and suppose Eq. (6.2) holds for all $k \in \{1, \ldots, n\}$, where $C_{n,k,r} \in \mathbb{Z}[X_2, \ldots, X_{n-k+1}]$, $k - 1 \leq r \leq n - 1$. Proposition 3.5 (ii) yields $S_{n+1,k} = T_{n,k}^{(a)} - T_{n,k}^{(b)} + T_{n,k}^{(c)}$ with

$$T_{n,k}^{(a)} = X_1 S_{n,k-1}, \quad T_{n,k}^{(b)} = (2n-1)X_2 S_{n,k}, \text{ and}$$
$$T_{n,k}^{(c)} = \sum_{j=1}^{n-k+1} X_1 X_{j+1} \frac{\partial S_{n,k}}{\partial X_j}.$$

By the induction hypothesis

$$T_{n,k}^{(a)} = X_1 \sum_{r=k-2}^{n-1} C_{n,k-1,r} X_1^r = \sum_{r=k-1}^{n} C_{n,k-1,r-1} X_1^r,$$

$$T_{n,k}^{(b)} = \sum_{r=k-1}^{n-1} (2n-1)X_2 C_{n,k,r} X_1^r,$$

$$T_{n,k}^{(c)} = X_1 X_2 \sum_{r=k-1}^{n-1} \frac{\partial}{\partial X_1}(C_{n,k,r} X_1^r)$$

---

[2] From the reciprocity law for Stirling polynomials, which is proved in Chapter II, one obtains a new and independent proof of Theorem 6.1, where no inductive argument is used. Cf. no. 27, Notes and supplements, Chapter II.

$$+ \sum_{j=2}^{n-k+1} \sum_{r=k-1}^{n-1} X_1 X_{j+1} \frac{\partial}{\partial X_j} (C_{n,k,r} X_1^r)$$

$$= \sum_{r=k-1}^{n-1} r X_2 C_{n,k,r} X_1^r + \sum_{r=k}^{n} \left(\sum_{j=2}^{n-k+1} X_{j+1} \frac{\partial C_{n,k,r-1}}{\partial X_j} \right) X_1^r.$$

For the desired expansion of $S_{n+1,k}$, we now have to find polynomials $C_{n+1,k,r} \in \mathbb{Z}[X_2, \ldots, X_{n-k+2}]$ that satisfy the recurrence

$$C_{n+1,k,r} = C_{n,k-1,r-1} - (2n-1-r) X_2 C_{n,k,r} + \sum_{j=2}^{n-k+1} X_{j+1} \frac{\partial C_{n,k,r-1}}{\partial X_j} \quad (6.3)$$

together with $C_{n,k,r} = 1$, if $n = k = r + 1$, and $C_{n,k,r} = 0$, if $k = 0$ or $k \le r = n - 1$ or $r < k - 1$. The polynomials $C_{n,k,r}$ are uniquely determined.

2. We set

$$C_{n,k,r} := (-1)^{n-1-r} \binom{2n-2-r}{k-1} \widetilde{B}_{2n-1-k-r,n-1-r}. \quad (6.4)$$

It suffices to show that $C_{n,k,r}$ meets the above conditions. First we check the initial values. Since $\widetilde{B}_{2n-1-k-r,n-1-r} = \widetilde{B}_{2(n-1-r)-l,n-1-r}$ where $l = k - 1 - r$, we get by Remark 5.5: $C_{n,k,r} = 0$ for $l \ge 1$, that is, for $r < k - 1$. The remaining cases are fairly clear.

Next we substitute (6.4) into Eq. (6.3). We start with the last summand on the right-hand side of Eq. (6.3) by first evaluating the partial derivatives:

$$\frac{\partial C_{n,k,r-1}}{\partial X_j} = (-1)^{n-r} \binom{2n-1-r}{k-1} \frac{\partial}{\partial X_j} \widetilde{B}_{2n-k-r,n-r}.$$

Corollary 4.4 now yields for $j \ge 2$

$$\frac{\partial}{\partial X_j} \widetilde{B}_{2n-k-r,n-r} = \binom{2n-k-r}{j} \widetilde{B}_{2n-k-r-j,n-r-1}.$$

Substituting the remaining C-terms into Eq. (6.3), we obtain after some straightforward calculation, in particular cancelling $(-1)^{n-r}$,

$$\left\{ \binom{2n-r}{k-1} - \binom{2n-1-r}{k-2} \right\} \widetilde{B}_{2n+1-k-r,n-r} =$$

$$(2n-1-r)\binom{2n-2-r}{k-1}X_2\widetilde{B}_{2n-1-k-r,n-1-r}$$

$$+\binom{2n-1-r}{k-1}\sum_{j=2}^{n-k+1}\binom{2n-k-r}{j}X_{j+1}\widetilde{B}_{2n-k-r-j,n-1-r}.$$

$$(6.5)$$

Using the identity

$$(2n-1-r)\binom{2n-2-r}{k-1}=(2n-k-r)\binom{2n-1-r}{k-1},$$

equation (6.5) is equivalently reduced to

$$\widetilde{B}_{2n+1-k-r,n-r}=\sum_{j=1}^{n-k+1}\binom{2n-k-r}{j}X_{j+1}\widetilde{B}_{2n-k-r-j,n-1-r},$$

which is the statement of Corollary 5.6. This completes the proof. $\Diamond$

Remark 6.1. (i) Taking $k=1$ on the right-hand side of the equation in Theorem 6.1, we obtain Comtet's expansion (1.7) (see [22, p. 151]). It should, however, be pointed out that Eq. (1.7) was intended for a solution of the Lagrange inversion problem (see Section 7 below). In this context, the idea of a connection between $\overline{f}_n$ in Eq. (1.7) and the Lie derivatives $(\theta D)^n$ studied by the same author [21] did not come into play.

(ii) Sylvester's note *On Reciprocants* (1886) deals with the task 'to express the successive derivatives of x in regard to y in terms of those of y in regard to x'. The solution is given in the form of the polynomial terms $S_{n,1}, 1 \le n \le 7$, probably one of the earliest times in the mathematical literature (cf. [98, p. 38] and [97]).

Theorem 6.1 enables us to find closed expressions for the coefficients $\sigma_{n,k}$ in Corollary 3.6 (i), thus leading to an explicit formula for $S_{n,k}$, which corresponds to that for the Bell polynomials in Proposition 4.3.

Proposition 6.2. *For every* $(r_1,\ldots,r_{n-k+1}) \in \mathbb{P}(2n-1-k,n-1)$ *we have*

$$\sigma_{n,k}(r_1,\ldots,r_{n-k+1})=(-1)^{n-1-r_1}\binom{2n-2-r_1}{k-1}\times$$

$$\times\beta_{2n-1-k-r_1,n-1-r_1}(0,r_2,\ldots)$$

$$=(-1)^{n-1-r_1}\beta(k-1,r_2,\ldots,r_{n-k+1}).$$

Proof. We first consider the coefficients of the associated Bell polynomials. $\widetilde{B}_{2n-1-k-r,n-1-r}$ is a sum over all partition types from $\mathbb{P}(2n - 1 - k - r, n - 1 - r)$, whose number r_1 of one-element blocks is zero. Let us abbreviate to $\widetilde{\mathbb{P}}_0$ the set of these 'associated' partition types. Then, according to Proposition 4.3 we can write

$$\widetilde{B}_{2n-1-k-r,n-1-r} = \sum_{\widetilde{\mathbb{P}}_0} \beta_{2n-1-k-r,n-1-r}(0, r_2, \ldots, r_{n-k+1}) \times$$

$$\times X_2^{r_2} \ldots X_{n-k+1}^{r_{n-k+1}}.$$

Observe now that any partition type $(r_1, \ldots, r_{n-k+1}) \in \mathbb{P}(2n-1-k, n-1)$ can be obtained by putting some $r = r_1 \in \{k-1, \ldots, n-1\}$ in the first place of $(0, r_2, \ldots, r_{n-k+1}) \in \widetilde{\mathbb{P}}_0$ (cf. Remark 5.3). This means that the double summation $\sum_{r=k-1}^{n-1} \sum_{\widetilde{\mathbb{P}}_0}$ in Theorem 6.1 can be replaced by $\sum_{\mathbb{P}(2n-1-k,n-1)}$. Comparing the coefficients of $X_1^{r_1} \cdot \ldots \cdot X_{n-k+1}^{r_{n-k+1}}$ in the resulting expression for $S_{n,k}$ and in that of Corollary 3.6 (i) then finally yields the desired $\sigma_{n,k}$. $\Diamond$

We now add to the list (1.10–1.12) in Section 1 a new mapping $\sigma : \mathbb{P} \longrightarrow \mathbb{Z}$, that might be called *Stirling function*, defined by

$$\sigma(r_1, r_2, r_3, \ldots) = (-1)^{n-1-r_1} \frac{(2n - 2 - r_1)!}{(k-1)! r_2! r_3! \cdot \ldots \cdot (2!)^{r_2} (3!)^{r_3} \cdot \ldots}. \quad (6.6)$$

Proposition 6.2 shows that $\sigma_{n,k}$ and σ agree on $\mathbb{P}(2n - 1 - k, n - 1)$.

Another succint expression of the relationship between β and σ is the following identity, which holds for all $(2n - 1 - k, n - 1)$-partition types.

Corollary 6.3.

$$\binom{2n-1-k}{r_1} \sigma(r_1, r_2, \ldots) = (-1)^{n-1-r_1} \binom{2n-2-r_1}{k-1} \beta(r_1, r_2, \ldots).$$

One could equivalently rewrite this equation as well for (n, k)-partition types as follows:

$$\binom{n}{r_1} \sigma(r_1, r_2, \ldots) = (-1)^{k-r_1} \binom{2k-r_1}{2k-n} \beta(r_1, r_2, \ldots).$$

Here, however, the assumption $2k \geq n$ must be satisfied (see Remark 5.3).

Example 6.1. The explicit form of $\sigma_{n,k}$ can be used to express $s_2(n,k)$ as a sum over $\mathbb{P}(2n-1-k, n-1)$. Suppose $\mathcal{F}$ is an extended function algebra. Then, from (2') in Examples 3.1 (i) we obtain $s_2(n,k) = A_{n,k}^{\log} \circ 1$. Corollary 3.6 (i) implies

$$A_{n,k}^{\log} = D(\log)^{-(2n-1)} \sum \sigma_{n,k}(r_1, r_2, \ldots) D^1(\log)^{r_1} D^2(\log)^{r_2} \cdots .$$

On the right-hand side we have $D^j(\log) = (-1)^{j-1}(j-1)! \cdot \mathrm{id}^{-j}$ for $1 \leq j \leq n-k+1$, and a straightforward calculation using Eq. (6.6) yields

$$s_2(n,k) = \sum_{\mathbb{P}(2n-1-k, n-1)} (-1)^{r_1-(k-1)} \frac{\binom{2n-2}{k-1}}{\binom{2n-2}{r_1}} \cdot \zeta(r_1, \ldots, r_{n-k+1}). \quad (6.7)$$

Remark 6.2. It looks like Corollary 6.3 expresses a kind of reciprocity law for the coefficient functions β and σ. However, it remains open whether there is a connection with the 'duality law' for the Stirling numbers[3]: $s_2(n,k) = c(-k, -n)$ (cf. [31, p. 25] and [50, p. 412]). In view of the fact that many MSP relations carry over more or less verbatim to Stirling numbers (see, e. g., Remark 4.1, Theorem 5.1, Corollary 5.2, Example 5.3), we could, in the reverse direction, try to establish a reciprocity law for the corresponding polynomials: $B_{n,k} = Z_{-k, -n}$. In the case $n \geq k \geq 0$ we define $Z_{n,k} := P_\zeta$ (see (1.11), the sum to be taken over $\mathbb{P}(n,k)$). The meaning of both $B_{n,k}$ and $Z_{n,k}$ can now be extended to arbitrary $n \in \mathbb{Z}$ with the aid of

$$B_{n,n-k} = \sum_{j=0}^{k} \binom{n}{k+j} X_1^{n-k-j} \widetilde{B}_{k+j,j},$$

which follows from Corollary 4.5 and also applies to the $Z_{n,n-k}$ (compare the corresponding statements concerning associative Stirling numbers in [36]; see also [18, ch. 8, exercises 10, 12]). Again we actually have $s_2(n,k) = B_{n,k}(1, \ldots, 1) = Z_{-k,-n}(1, \ldots, 1) = c(-k, -n)$. However, $B_{n,k} = Z_{-k,-n}$ in general turns out to be false.

Counterexample: $B_{5,2} \neq Z_{-2,-5} = 105 X_1^{-8} X_2^3 - 120 X_1^{-7} X_2 X_3 + 30 X_1^{-6} X_4$, whereas $B_{5,2}(1,1,1) = Z_{-2,-5}(1,1,1) = 15$.

Equation (i) of the theorem to follow is the MSP version of a well-known formula (cf. Example 6.2 below) expressing the signed Stirling numbers of the first kind in terms of the Stirling numbers of the second kind.

---

[3] Chapter II answers this question in the affirmative.

Theorem 6.4 (Schlömilch type formulas). *For* $n \geq k \geq 1$

(i) $\quad A_{n,k} = \sum_{r=k-1}^{n-1} (-1)^{n-1-r} \binom{2n-2-r}{k-1} \binom{2n-k}{r+1-k} \times$

$$\times X_1^{r-2n+1} B_{2n-1-k-r,n-1-r},$$

(ii) $\quad B_{n,k} = \sum_{r=k-1}^{n-1} (-1)^{n-1-r} \binom{2n-2-r}{k-1} \binom{2n-k}{r+1-k} \times$

$$\times X_1^{2n-1-r} A_{2n-1-k-r,n-1-r}.$$

Proof. For the purpose of formal convenience we adhere to the convention $B_{i,j} = 0$ (and consequently $\widetilde{B}_{i,j} = 0$), if $j < 0$. Then, the equation of Corollary 4.5 can be rewritten in the form

$$B_{n,k} = \sum_{j=0}^{n} \binom{n}{j} X_1^{n-j} \widetilde{B}_{j,k-(n-j)},$$

whence by binomial inversion

$$\widetilde{B}_{n,k} = \sum_{j=0}^{k} (-1)^j \binom{n}{j} X_1^j B_{n-j,k-j}. \tag{6.8}$$

The upper limit n has been replaced here by k, since we have $B_{n-j,k-j} = 0$ for $j > k$. Substituting Eq. (6.8) into the equation of Theorem 6.1 gives

$$A_{n,k} = X_1^{-(2n-1)} S_{n,k}$$

$$= \sum_{r=k-1}^{n-1} \sum_{j=0}^{n-1-r} (-1)^{n-1-r-j} \binom{2n-2-r}{k-1} \binom{2n-1-k-r}{j} \times$$

$$\times X_1^{r-2n+1+j} B_{2n-1-k-r-j,n-1-r-j}.$$

Define now a new index $s = r + j$ so that $k - 1 \leq s \leq n - 1$. Interchanging the order of summation then leads to

$$A_{n,k} = \sum_{s=k-1}^{n-1} (-1)^{n-1-s} X_1^{s-2n+1} B_{2n-1-k-s,n-1-s} \times \cdots$$

$$\cdots \times \underbrace{\sum_{r=k-1}^{s} \binom{2n-2-r}{k-1}\binom{2n-1-k-r}{s-r}}_{(*)}.$$

The sum $(*)$ can be simplified by using elementary properties of the binomial coefficients. First check that its summand is equal to $\binom{2n-2-r}{2n-2-s}\binom{2n-2-s}{k-1}$. Then, after some calculation, $(*)$ is reduced to

$$(*) = \binom{2n-2-s}{k-1}\sum_{r=k-1}^{s}\binom{2n-2-r}{2n-2-s} = \binom{2n-2-s}{k-1}\binom{2n-k}{s+1-k}.$$

Finally, renaming the index s to r gives the assertion (i).

Now we derive (ii) from (i). Using Proposition 3.4 and applying $P \longmapsto P^{\overline{\varphi}}$ to both sides of (i) makes the left-hand side into $B_{n,k}^{\varphi} \circ \overline{\varphi}$, while the term $X_1^{r-2n+1} B_{2n-1-k-r,n-1-r}$ on the right becomes $D(\overline{\varphi})^{r-2n+1} B_{2n-1-k-r,n-1-r}^{\overline{\varphi}}$. Next, we apply $\circ \varphi$ (from the right) to both sides of (i). It follows

$$B_{n,k}^{\varphi} = \sum_{r=k-1}^{n-1} (-1)^{n-1-r}\binom{2n-2-r}{k-1}\binom{2n-k}{r+1-k} \times$$

$$\times (D(\overline{\varphi})^{r-2n+1} \circ \varphi) \cdot (B_{2n-1-k-r,n-1-r}^{\overline{\varphi}} \circ \varphi).$$

Observe now that $D(\overline{\varphi})^{r-2n+1} \circ \varphi = D(\varphi)^{2n-1-r} = (X_1^{2n-1-r})^{\varphi}$ and, again by Proposition 3.4, $B_{2n-1-k-r,n-1-r}^{\overline{\varphi}} \circ \varphi = A_{2n-1-k-r,n-1-r}^{\varphi}$. This completes the proof. (The argument amounts to directly applying Theorem 5.3 (ii), that is, replacing each X_j with $A_{j,1}$ on both sides of (i).) $\Diamond$

Example 6.2. Specializing Theorem 6.4 (i) by taking all indeterminates to 1, we immediately obtain *Schlömilch's formula for the Stirling numbers of the first kind* (see e. g. [18]):

$$s_1(n,k) = \sum_{r=k-1}^{n-1} (-1)^{n-1-r}\binom{2n-2-r}{k-1}\binom{2n-k}{r+1-k} \times \qquad (6.9)$$

$$\times s_2(2n-1-k-r,n-1-r).$$

Likewise we get from Theorem 6.1 a slightly shorter formula of a similar type:

$$s_1(n,k) = \sum_{r=k-1}^{n-1} (-1)^{n-1-r} \binom{2n-2-r}{k-1} \tilde{s}_2(2n-1-k-r, n-1-r).$$

(6.10)

Eq. (6.10) runs with smaller numbers than does (6.9). Compare, for instance, the computation of $s_1(7,4) = -84 \cdot 90 + 56 \cdot 150 - 35 \cdot 45 = -735$ by Schlömilch's formula (6.9) with that of $s_1(7,4) = -84 \cdot 15 + 56 \cdot 10 - 35 \cdot 1 = -735$ by Eq. (6.10). However, associated Stirling numbers of the second kind do not have quite simple explicit representations (cf. [36], in particular equation (3.11), ibid.).

Finally it should be noted that the statement (ii) of Theorem 6.4 enables the Stirling numbers of the second kind to be represented using the Stirling numbers of the first kind. This result is due to Gould [32].

7 Remarks on Lagrange inversion

Let φ be an analytic function (bijective in the real or complex domain), which is given in the form of a power series

$$\varphi(x) = \sum_{n \geq 1} f_n \frac{x^n}{n!}$$

with zero constant term and $f_1 \neq 0$. Then, the compositional inverse $\overline{\varphi}$ is unique, and one may ask for the coefficients $\overline{f}_n$ in the series expansion $\overline{\varphi}(x) = \sum_{n \geq 1} \overline{f}_n x^n / n!$. One possible answer to this is Lagrange's famous inversion formula (cf. [22] and [95] for proofs and further references):

$$\overline{f}_n = \left(\frac{d}{dx}\right)^{n-1} \left(\left(\frac{x}{\varphi(x)}\right)^n\right)\Bigg|_{x=0}.$$

(7.1)

This innocent looking expression, however, provides in most cases an all but simple method of computing. Many attempts have therefore been made to obtain alternative and more efficient expressions (see, e. g., [54], [99]). Trying to express $\overline{f}_n$ as a function of $f_1, f_2, \ldots, f_n$ is a quite natural approach. Morse and Feshbach [71, p. 412], for example, employed the Residue Theorem to show that $\bar{f}_n$ can be represented as a polynomial expression over all partitions of $n - 1$. Comtet [22] derived the remarkable result that the right-hand side of Eq. (7.1) is equal to $\sum_{k=0}^{n-1}(-1)^k f_1^{-n-k} B_{n+k-1,k}(0, f_2, \ldots, f_n)$. By Theorem 6.1 the latter can easily be seen to agree with $A_{n,1}(f_1, \ldots, f_n)$, that is, we have for all $n \geq 1$:

$$\bar{f}_n = A_{n,1}(f_1, \ldots, f_n). \tag{7.2}$$

In the following we will deal with some few function-algebraic aspects of Lagrange inversion. It turns out that Eq. (7.2) can be proved without using (7.1). We will also see that inverting a function φ is nothing else than switching from $D_\iota^n(\varphi)(0)$ to $D_\varphi^n(\iota)(0)$ in the corresponding series expansions. A few examples should briefly illustrate the computational aspects.

Let $\mathcal{F}$ be an extended function algebra, $\varphi \in \mathcal{F}$. Throughout this section we assume $f_0 := \varphi \circ 0 = 0$, and $f_1 := D(\varphi) \circ 0$ to be a unit.[4] Since the notion of convergence has no place in $\mathcal{F}$, we make use of formal power series.

Definition 7.1. We say that φ is an *exponential generating function* of the sequence of constants $f_0, f_1, f_2, \ldots \in \mathcal{K}$ (symbolically written $\varphi(x) = \sum_{n \geq 0} f_n x^n / n!$), if $D^n(\varphi)(0) = f_n$ for every $n \geq 0$.[5]

The exponential generating function of the constant sequence $1, 1, 1, \ldots$

$$e^x := \exp(x) := \sum_{n \geq 0} \frac{x^n}{n!}$$

obviously has the properties that one would expect from an exponential (as has been indicated in Section 2.2).

Our basic statement on inversion is now the following

Proposition 7.1.

$$\text{(i)} \quad \varphi(x) = \sum_{n \geq 1} D_{\text{id}}^n(\varphi)(0) \frac{x^n}{n!} = \sum_{n \geq 1} B_{n,1}^\varphi(0) \frac{x^n}{n!},$$

$$\text{(ii)} \quad \bar{\varphi}(x) = \sum_{n \geq 1} D_\varphi^n(\text{id})(0) \frac{x^n}{n!} = \sum_{n \geq 1} A_{n,1}^\varphi(0) \frac{x^n}{n!}.$$

---

[4] In the sequel we use the customary notation $g(0)$ instead of $g \circ 0$, $g \in \mathcal{F}$
[5] In the case $f_0 = 0$ we take the lower summation limit to 1.

Proof. (i): Clearly $D_{\mathrm{id}} = D$. Hence for all $n \geq 1$

$$B_{n,1}^{\varphi}(0) = (X_n)^{\varphi}(0) = D^n(\varphi)(0) = D_{\mathrm{id}}^n(\varphi)(0).$$

(ii): By Proposition 3.4 (ii) we have

$$A_{n,1}^{\varphi}(0) = (B_{n,1}^{\overline{\varphi}} \circ \varphi)(0) = (D^n(\overline{\varphi}) \circ \varphi)(0).$$

Thus Pourchet's formula yields

$$D_{\varphi}^n(\mathrm{id})(0) = (D^n(\overline{\varphi}) \circ \varphi)(0) = D^n(\overline{\varphi})(\varphi(0)) = D^n(\overline{\varphi})(0). \qquad \Diamond$$

Corollary 7.2. *For every* $n \geq 1$

$$\overline{f}_n = A_{n,1}(f_1, \ldots, f_n)$$

$$= \frac{1}{f_1^{2n-1}} \cdot \sum_{\mathbb{P}(2n-2,n-1)} \frac{(-1)^{n-1-r_1} \cdot (2n-2-r_1)!}{r_2! \cdots r_n! \cdot (2!)^{r_2} \cdots (n!)^{r_n}} f_1^{r_1} f_2^{r_2} \cdots f_n^{r_n}.$$

Proof. Proposition 7.1 (ii) yields $\overline{f}_n = A_{n,1}^{\varphi}(0) = A_{n,1}(f_1, \ldots, f_n)$, where $f_j = D^j(\varphi)(0)$ $(j = 1, \ldots, n)$. Then, applying Corollary 3.6 (i) and Eq. (6.6) for $k = 1$ gives the asserted. $\qquad \Diamond$

Except for some simple cases, the higher Lie derivatives $D_{\varphi}^n(\mathrm{id})$ turn out to be considerably less complex than the Lagrangian terms $D^{n-1}((\iota/\varphi)^n)$. Nonetheless, the most advantage is presumably to be gained from applying Theorem 6.1 (with $k = 1$) or Corollary 7.2 to the coefficients $f_j = D^j(\varphi)(0)$ $(j = 1, 2, \ldots, n)$.

Let us consider three examples.

Example 7.1. Let $\varphi(x) = xe^{-x}$. As is well-known (see, e.g., [95, p. 23]), the inverse $\overline{\varphi}$ is exponential generating function of the sequence $r(n)$ $(n = 1, 2, 3, \ldots)$ of numbers of rooted (labeled) trees on n vertices. The function φ seems to be tailored to the application of Eq. (7.1), for we have $D^{n-1}((\mathrm{id}/\varphi)^n)(x) = D^{n-1}(\exp \circ (n \cdot \mathrm{id}))(x) = n^{n-1}e^{nx}$ and thus readily by Eq. (7.1): $r(n) = \overline{f}_n = n^{n-1}e^{nx}|_{x=0} = n^{n-1}$.

Compared with this neat calculation, applying $A_{n,1}$ to the coefficients $f_j = D^j(\varphi)(0) = (-1)^{j-1}j$ $(j = 1, \ldots, n)$ gives less satisfactory results. Corollary 7.2 yields

$$\sum_{\mathbb{P}(2n-2,n-1)} (-1)^{r_1} \frac{(2n-2-r_1)!}{r_2! r_3! \cdots r_n!} \cdot \frac{1}{1!^{r_2} 2!^{r_3} \cdots (n-1)!^{r_n}}. \qquad (7.3)$$

Here a combinatorial argument is needed to see that (7.3) equals n^{n-1}.

Example 7.2. Let $\varphi(x) = e^x - 1$. While formula (7.1) fails to yield simple expressions, the Taylor series expansion of $\overline{\varphi} = \log \circ (1 + \mathrm{id})$ is immediately obtained by Proposition 7.1 (ii). We have $A_{n,1}^{\varphi}(0) = A_{n,1}(1, \ldots, 1) = s_1(n, 1) = (-1)^{n-1}(n-1)!$ (see Examples 3.1 (iii)), and hence $\log(1 + x) = \sum_{n \geq 1} (-1)^{n-1} x^n / n$.

Example 7.3. Let $\varphi(x) = 1 + 2x - e^x$. According to Stanley [95, p. 13], $\overline{\varphi}$ is exponential generating function of the sequence $t(n)$ $(n = 1, 2, 3, \ldots)$, where $t(n)$ denotes the number of total partitions of $\{1, \ldots, n\}$; cf. the fourth problem posed by E. Schröder (1870) ('arbitrary set bracketings') [95, p. 178]. When applied to φ, the Lagrange formula does not 'seem to yield anything interesting' (Stanley). Let us therefore have a look at what can be achieved with the help of the MSPs.

We have $f_1 = 1, f_j = -1$ $(j \geq 2)$, and hence by Corollaries 6.3 and 7.2

$$t(n) = A_{n,1}(1, -1, \ldots, -1) = \sum_{\mathbb{P}(2n-2,n-1)} \binom{2n-2}{r_1}^{-1} \beta(r_1, \ldots, r_n).$$

The same can be expressed as well in terms of associated Stirling numbers of the second kind by applying Theorem 6.1:

$$t(n) = \sum_{r=0}^{n-1} \tilde{s}_2(2n - 2 - r, n - 1 - r).$$

Regarding the latter formula, the reader is referred to Comtet [22, p. 224].

Alternatively, we can use the fact that $\overline{f}_n = D_{\varphi}^n(\mathrm{id})(0)$ (cf. Proposition 7.1 (ii)). This leads to a recursive solution. The repeated application of the Lie derivation D_{φ} inductively results in a representation of the form

$$D_{\varphi}^n(\mathrm{id})(x) = (2 - e^x)^{-(2n-1)} T_n(x), \quad \text{where}$$

$$T_1(x) = 1, \quad T_n(x) = \sum_{k=0}^{n-1} b_{n-1,k} e^{kx} \quad \text{for } n \geq 2,$$

with non-negative integers $b_{n-1,k}$. It follows for all $n \geq 1$

$$t(n) = T_n(0) = b_{n-1,0} + b_{n-1,1} + \cdots + b_{n-1,n-1}.$$

Equating now the coefficients of e^{kx} in $D_\varphi^{n+1}(\mathrm{id})(x) = D_\varphi(D_\varphi^n(\mathrm{id}))(x)$ gives the recurrence

$$b_{n,k} = (2n - k)b_{n-1,k-1} + 2kb_{n-1,k} \qquad (1 \le k \le n),$$
$$b_{i,0} = \delta_{i0}, \quad b_{i,j} = 0 \qquad (0 \le i < j).$$

Some special values:

$$b_{n,1} = 2^{n-1}, \quad b_{n,2} = 2^{n-1}(2^n - n - 1), \quad b_{n,n} = n!.$$

We obtain $t(1) = 1$, $t(2) = 0 + 1 = 1$, $t(3) = 0 + 2 + 2 = 4$, $t(4) = 0 + 2^2 + 2^2(2^3 - 3 - 1) + 3! = 26$.

Our last statement concerns exponential generating functions of the form $\exp \circ (t \cdot \varphi)$, $t \in \mathcal{K}$.

Proposition 7.3.

$$\text{(i)} \quad e^{t\varphi(x)} = \sum_{n \ge 0} \left(\sum_{k=0}^{n} B_{n,k}^\varphi(0) t^k \right) \frac{x^n}{n!},$$

$$\text{(ii)} \quad e^{t\overline{\varphi}(x)} = \sum_{n \ge 0} \left(\sum_{k=0}^{n} A_{n,k}^\varphi(0) t^k \right) \frac{x^n}{n!}.$$

Proof. (i): By Faà di Bruno's formula (4.1) and Remark 2.2

$$D^n(e^{t\varphi}) = \sum_{k=0}^{n} B_{n,k}^{t\varphi} \cdot (D^k(\exp) \circ (t\varphi)) = e^{t\varphi} \sum_{k=0}^{n} B_{n,k}^\varphi t^k,$$

hence

$$D^n(e^{t\varphi})(0) = e^{t\varphi(0)} \sum_{k=0}^{n} B_{n,k}^\varphi(0) \cdot (t \circ 0)^k.$$

Since $e^{t\varphi(0)} = e^{t \cdot 0} = e^0 = 1$ and $t \circ 0 = t$, we are done.

Note $\overline{\varphi}(0) = 0$; then (ii) follows from (i) by Proposition 3.4. ◇

We conclude with a well-known example.

Example 7.4. As in Example 7.2, take $\varphi = \exp -1$. From (2) in Examples 3.1 (i) we have $B^{\varphi}_{n,k}(0) = s_2(n,k)$. Then by Proposition 7.3 (i)

$$\exp(t(e^x - 1)) = \sum_{n \geq 0} \left(\sum_{k=0}^{n} s_2(n,k)\, t^k \right) \frac{x^n}{n!}.$$

Put $t = 1$. This shows that $\exp(e^x - 1)$ is exponential generating function of the Bell number sequence $b(n) := \sum_{k=0}^{n} s_2(n,k)$ (see [95, p. 13]).

We also have $A^{\varphi}_{n,k}(0) = s_1(n,k)$ (from (1) in Examples 3.1 (i)). Since $\overline{\varphi} = \log \circ (1 + \iota)$, Proposition 7.3 (ii) yields

$$\exp(t \log(1 + x)) = \sum_{n \geq 0} \left(\sum_{k=0}^{n} s_1(n,k)\, t^k \right) \frac{x^n}{n!}. \tag{7.4}$$

Thus $(1 + x)^t$ turns out to be the exponential generating function of the sequence $\sum_{k=0}^{n} s_1(n,k)\, t^k$ $(n = 1, 2, 3, \ldots)$ (see, e. g., [18, p. 281]).

8 Concluding remarks

The preceding work was primarily intended as an attempt to introduce, within a general function-algebraic setting, the notion of MSP of the first and second kind (the latter being identical to that of partial Bell polynomial). The investigation was focussed on establishing the inverse relationship as well as other fundamental properties of the two polynomial families. The resulting picture shows that the MSPs may be understood as a kind of strong generalizations of the corresponding Stirling numbers.

Supplementary to this the reader is referred to a package [82] for *Mathematica*® providing functions that generate the expressions $S_{n,k}$, $A_{n,k}$, and $B_{n,k}$ together with a substitution mechanism.

No attempt was made here to develop a combinatorial interpretation of the coefficient function $\sigma : \mathbb{P} \longrightarrow \mathbb{Z}$ (in $S_{n,k}$ and $A_{n,k}$). Based on [80], Haiman and Schmitt [33] have offered a satisfactory explanation of Comtet's expansion (1.7) (cf. Remark 6.1) from an incidence algebra point of view (using colored partitions of finite colored sets). So, it might be worth-while examining whether this idea will also apply to the general case formulated in Theorem 6.1. The 'ban on one-element blocks' observed in the case $k = 1$ [33, p. 180] may be seen as a sign in this direction, since it is as well a striking feature of the whole family $S_{n,k}$ (see, e. g., Remark 5.3, Proposition 6.2, and Eq. (6.6)).

NOTES AND SUPPLEMENTS

1. Product rule [p. 15].   Suppose that $\mathcal{F}$ is a ring with char $\mathcal{F} \neq 2$ and $D : \mathcal{F} \longrightarrow \mathcal{F}$ an additive mapping with $D(f^2) = 2fD(f)$ for all $f \in \mathcal{F}$. Then $D(f \cdot g) = D(f) \cdot g + f \cdot D(g)$ for all $f, g \in \mathcal{F}$. (Exercise)

2. Logarithmic derivative identity [p. 15].   Let $f_1, \ldots, f_n$ be units of $\mathcal{F}$ and $j_1, \ldots, j_n \in \mathbb{Z}$. Then

$$\frac{D(f_1^{j_1} \cdots f_n^{j_n})}{f_1^{j_1} \cdots f_n^{j_n}} = \sum_{k=1}^{n} j_k \frac{D(f_k)}{f_k}.$$

This can be inferred from Eq. (2.1). (Exercise)

3. Notion of constant [p. 15].   Let $\mathcal{F}$ be a function algebra. According to Menger's definition in [64], a function f is a *constant* if $f \circ 0 = f$. Let f be constant and g be any function from $\mathcal{F}$. Then we have $f \circ g = (f \circ 0) \circ g = f \circ (0 \circ g) = f \circ 0 = f$.

   In a function algebra with derivation D it follows from the chain rule (D3): $D(f) = D(f \circ 0) = 0$, that is, every constant of $\mathcal{F}$ is also a constant of the differential ring $(\mathcal{F}, D)$. The question of whether $D(f) = 0$ implies $f \circ 0 = f$ remains open. Axiom (D5) ensures that $\mathcal{F}$ and $(\mathcal{F}, D)$ have the same constants.

4. Generalized chain rule [p. 16].   *Proof* of Eq. (2.3): The polynomial P can be written in the form

$$P = \sum \alpha(r_1, \ldots, r_n) X_1^{r_1} \cdots X_n^{r_n},$$

where the sum ranges over a finite set of sequences $(r_1, \ldots, r_n)$ of non-negative integers, and $\alpha(r_1, \ldots, r_n) \in \mathcal{K}$. Recall that for j with $1 \leq j \leq n$

$$\frac{\partial P}{\partial X_j} = \sum r_j \alpha(r_1, \ldots, r_n) X_1^{r_1} \cdots X_j^{r_j - 1} \cdots X_n^{r_n}.$$

Now, given any functions $f_1, \ldots, f_n \in \mathcal{F}$, we obtain

$$D(P(f_1, \ldots, f_n)) = \sum \alpha(r_1, \ldots, r_n) D(f_1^{r_1} \cdots f_n^{r_n})$$

$$= \sum \alpha(r_1, \ldots, r_n) \sum_{k=1}^{n} f_1^{r_1} \cdots D(f_k^{r_k}) \cdots f_n^{r_n}$$

$$= \sum_{k=1}^{n} \sum r_k \alpha(r_1, \ldots, r_n) f_1^{r_1} \cdots f_k^{r_k-1} \cdots f_n^{r_n} \cdot D(f_k)$$

$$= \sum_{k=1}^{n} \frac{\partial P}{\partial X_k}(f_1, \ldots, f_n) \cdot D(f_k).$$

5. Homogeneous polynomials [p. 16]. The polynomial P (from no. 4) is said to be *homogeneous of degree* k, if $r_1 + \cdots + r_n = k$ for every n-tuple $(r_1, \ldots, r_n)$ from its index set. Euler's homogeneous function theorem also applies to homogeneous polynomials $P \in \mathcal{K}[X_1, \ldots, X_n]$ of degree k. The following holds:

$$\sum_{j=1}^{n} X_j \frac{\partial P}{\partial X_j} = k \cdot P.$$

(Exercise)

6. Generalized inversion rule [p. 20]. Let $f \in \mathcal{F}$ be any function whose inverse exists, and suppose that $D(f)$ and $D(f) \circ \overline{f}$ are units. Then for every integer $m \geq 0$: $D(\overline{f})^m \circ f = D(f)^{-m}$.

Proof (by induction). For $m = 0$ the assertion becomes $1 \circ f = 1$ (true because 1 is a constant). Fix an arbitrary $m \geq 0$ and assume (induction hypothesis) $D(\overline{f})^m \circ f = D(f)^{-m}$. Then by (F3) and Eq. (2.2) $D(\overline{f})^{m+1} \circ f = (D(\overline{f})^m \cdot D(\overline{f})) \circ f = (D(\overline{f})^m \circ f) \cdot (D(\overline{f}) \circ f) = D(f)^{-m} \cdot D(f)^{-1} = D(f)^{-(m+1)}$.

7. Recurrences [p. 21]. Let k, n be integers with $1 \leq k \leq n$. Then the following holds:

$$B_{n,n-k} = \sum_{j=0}^{n-k-1} \sum_{r=1}^{k} \binom{n-j-1}{r} X_1^j X_{r+1} B_{n-j-r-1,n-j-k-1}. \qquad (*)$$

Proof (Outline). From Proposition 3.5 (iii) follows

$$B_{n,n-k} = X_1 B_{n-1,n-k-1} + \sum_{j=1}^{k} X_{j+1} \frac{\partial B_{n-1,n-k}}{\partial X_j},$$

whence by an inductive argument

$$B_{n,n-k} = \sum_{j=0}^{n-k-1} X_1^j F(n-j,k) \quad \text{with}$$

$$F(n-j,k) := \sum_{r=1}^{k} X_{r+1} \frac{\partial B_{n-j-1,n-j-k}}{\partial X_r}.$$

By Corollary 4.4 we have

$$\frac{\partial B_{n-j-1,n-j-k}}{\partial X_r} = \binom{n-j-1}{r} B_{n-j-r-1,n-j-k-1}$$

thus obtaining the assertion. $\diamondsuit$

Applying Theorem 5.3 (i) to (*) immediately yields

$$A_{n,n-k} = \sum_{j=0}^{n-k-1} \sum_{r=1}^{k} \binom{n-j-1}{r} X_1^{-j} A_{r+1,1} A_{n-j-r-1,n-j-k-1}.$$

8. Normal ordering [p. 24].   The expansion of D_φ^n ($n \geq 0$) into a finite linear combination of $D^0, D^1, D^2, \ldots, D^n$ (as stated in Proposition 3.1) can be viewed as the normal ordering of the ›word‹ $(D(\varphi)^{-1}D)^n$ that brings about a »form where all operators D stand to the right« (cf. Mansour and Schork [59, p. xvii]). Equation (1.1) is one of the oldest results in this field and can be traced back to H. F. Scherk (1823) (cf. ibid., p. 13). We readily obtain it from Proposition 3.1 by setting $\varphi = \log$ and observing that $A_{n,k}^{\log} = s_2(n,k)\iota^k$ (see Examples 3.1 (i),(2')). By interpreting xD as $D_{\log}$ the Stirling numbers of the second kind get associated to the logarithm. It therefore fits well into the picture that the signed Stirling numbers of the first kind arise from the normal ordering of $(e^{-x}D)^n$ (or, equivalently, from expanding the differential operator $D_{\exp}^n$):

$$(e^{-x}D)^n = D_{\exp}^n = e^{-nx} \sum_{k=0}^{n} s_1(n,k)D^k.$$

In their treatise [59] Mansour and Schork give »an introduction to combinatorial aspects of normal ordering in the Weyl algebra and some of its close relatives«. In addition, some interesting connections that the topic has with quantum theory are discussed.

9. McKiernan's chain rule [p. 25].   Just as in Menger's tri-operational algebra of analysis [64], the chain rule is one of the defining axioms in a function algebra with derivation $(\mathcal{F}, D)$ (in the sense of Definition 2.1). This results in the problem of finding in $\mathcal{F}$ an expression for the nth derivative of a composite function. In his commentary on Menger's algebra Sklar [92] states: »Several equivalent forms of such a formula exist, of which the oldest and best-known is that of Francesco Faà di Bruno [...] But the most perspicuous form is surely that developed by M. A. McKiernan, who had been a student of Menger's.« It appears in Mengerian notation as

$$D^n(f \circ g) = \sum_{k=0}^{n} \left((D^k(f) \circ g) \cdot \sum_{j=0}^{k} \frac{(-1)^{k-j}}{j!(k-j)!} g^{k-j} D^n(g^j) \right).$$

However, this formula obviously has the same structure as that of Faà di Bruno (see Eq. (4.1)), and the latter results directly in McKiernan's expression by applying Bertrand's formula (Proposition 4.2) to $B_{n,k}^g$ in Eq. (4.1).

10. Table: Partial Bell polynomials [p. 27].

a) The partial Bell polynomials $B_{7,k}$ ($1 \le k \le 7$):

$$B_{7,1} = X_7$$
$$B_{7,2} = 35X_3X_4 + 21X_2X_5 + 7X_1X_6$$
$$B_{7,3} = 21X_5X_1^2 + 70X_3^2X_1 + 105X_2X_4X_1 + 105X_2^2X_3$$
$$B_{7,4} = 35X_4X_1^3 + 210X_2X_3X_1^2 + 105X_2^3X_1$$
$$B_{7,5} = 35X_3X_1^4 + 105X_2^2X_1^3$$
$$B_{7,6} = 21X_1^5X_2$$
$$B_{7,7} = X_1^7$$

b) The partial Bell polynomials $B_{n,n-k}$ ($k = 0, 1, 2, 3$):

$$B_{n,n} = X_1^n$$
$$B_{n,n-1} = \binom{n}{2} X_1^{n-2}X_2$$
$$B_{n,n-2} = 3\binom{n}{4} X_1^{n-4}X_2^2 + \binom{n}{3} X_1^{n-3}X_3$$
$$B_{n,n-3} = \binom{n}{4} X_1^{n-4}X_4 + 10\binom{n}{5} X_1^{n-5}X_2X_3 + 15\binom{n}{6} X_1^{n-6}X_2^3$$

11. A consequence from homogeneity [p. 27]. For $1 \le k \le n$ the recurrence holds

$$B_{n,k} = \frac{1}{k} \sum_{j=1}^{n-k+1} \binom{n}{j} X_j B_{n-j,k-1}.$$

This results from no. 5 and Corollary 4.4. Derive the corresponding identity for Stirling numbers of the second kind. (Exercise)

12. Table: Associated Bell polynomials [p. 28].

$$\widetilde{B}_{10,1} = X_{10}$$
$$\widetilde{B}_{10,2} = 126X_5^2 + 210X_4X_6 + 120X_3X_7 + 45X_2X_8$$
$$\widetilde{B}_{10,3} = 630X_6X_2^2 + 1575X_4^2X_2 + 2520X_3X_5X_2 + 2100X_3^2X_4$$
$$\widetilde{B}_{10,4} = 3150X_4X_2^3 + 6300X_3^2X_2^2$$
$$\widetilde{B}_{10,5} = 945X_2^5$$

We have in general: $\widetilde{B}_{n,k} = 0$ for $k > \left[\frac{n}{2}\right]$. (Exercise)

13. Unsigned Lah numbers [p. 28]. Let $(r_1, r_2, \ldots) \in \mathbb{P}(n, k)$. Then, the order function (1.10) can be written

$$\omega(r_1, r_2, \ldots) = \frac{n!}{r_1!r_2! \cdots r_{n-k+1}!} = \frac{n!}{k!} \cdot \frac{k!}{r_1!r_2! \cdots r_{n-k+1}!}.$$

On the other hand

$$\sum_{\mathbb{P}(n,k)} \frac{k!}{r_1!r_2! \cdots r_{n-k+1}!} = \binom{n-1}{k-1}. \qquad (*)$$

Exercise: Find a bijective proof for (*).

This gives us a combinatorial interpretation and representation of the (unsigned) Lah numbers:

$$l^+(n,k) := P_\omega(1, \ldots, 1) = \sum_{\mathbb{P}(n,k)} \omega(r_1, r_2, \ldots) = \frac{n!}{k!}\binom{n-1}{k-1}.$$

14. Table: Unsigned Lah polynomials [p. 28]. The 7th generation of the unsigned Lah polynomials:

$$L_{7,1}^+ = 5040X_7$$

$$L_{7,2}^+ = 5040X_1X_6 + 5040X_3X_4 + 5040X_2X_5$$
$$L_{7,3}^+ = 2520X_1^2X_5 + 2520X_1X_3^2 + 5040X_1X_2X_4 + 2520X_2^2X_3$$
$$L_{7,4}^+ = 840X_1^3X_4 + 2520X_1^2X_2X_3 + 840X_1X_2^3$$
$$L_{7,5}^+ = 210X_1^4X_3 + 420X_1^3X_2^2$$
$$L_{7,6}^+ = 42X_1^5X_2$$
$$L_{7,7}^+ = X_1^7$$

15. Signed Lah numbers [p. 31].   The signed Lah numbers $l(n,k)$ can be obtained as $A_{n,k}^\varphi(0) = B_{n,k}^\varphi(0) =: l(n,k) = (-1)^n l^+(n,k)$, where $\varphi(x) = -x/(1+x)$ (see Examples 5.1 (ii)).   This implies the well-known identity, which expresses the self-inversion of the signed Lah numbers:

$$\sum_{j=k}^{n} l(n,j)l(j,k) = \delta_{nk} \quad (1 \le k \le n).$$

As a consequence, one has

$$\sum_{j=k}^{n} (-1)^{n-j} \binom{n}{j}\binom{j}{k} = \delta_{nk} \quad (1 \le k \le n).$$

This identity (*binomal inversion*) often appears in the literature as the equivalence

$$\forall n \ge 0: b_n = \sum_{k=0}^{n}(-1)^{n-k}\binom{n}{k}a_k \iff \forall n \ge 0: a_n = \sum_{k=0}^{n}\binom{n}{k}b_k.$$

(Exercise)

16. Table: Stirling polynomials [p. 38].   a) According to Theorem 5.1 the (multivariate) Stirling polynomials $A_{n,k}$ of the first kind are inverse (orthogonal) companions of the partial Bell polynomials (= Stirling polynomials of the second kind). To be precise, they actually are Laurent polynomials. The following list comprises their 6th generation:

$$A_{6,1} = -\frac{945X_2^5}{X_1^{11}} + \frac{1260X_3X_2^3}{X_1^{10}} - \frac{210X_4X_2^2}{X_1^9} - \frac{280X_3^2X_2}{X_1^9}$$
$$+ \frac{21X_5X_2}{X_1^8} + \frac{35X_3X_4}{X_1^8} - \frac{X_6}{X_1^7}$$

$$A_{6,2} = \frac{945X_2^4}{X_1^{10}} - \frac{840X_3X_2^2}{X_1^9} + \frac{105X_4X_2}{X_1^8} + \frac{70X_3^2}{X_1^8} - \frac{6X_5}{X_1^7}$$

$$A_{6,3} = -\frac{420X_2^3}{X_1^9} + \frac{210X_3X_2}{X_1^8} - \frac{15X_4}{X_1^7}$$

$$A_{6,4} = \frac{105X_2^2}{X_1^8} - \frac{20X_3}{X_1^7}$$

$$A_{6,5} = -\frac{15X_2}{X_1^7}$$

$$A_{6,6} = \frac{1}{X_1^6}$$

b) The multivariate Stirling polynomials $A_{n,n-k}$ $(k = 0, 1, 2, 3, 4)$:

$$A_{n,n} = \frac{1}{X_1^n}$$

$$A_{n,n-1} = -\binom{n}{2}\frac{X_2}{X_1^{n+1}}$$

$$A_{n,n-2} = 3\binom{n+1}{4}\frac{X_2^2}{X_1^{n+2}} - \binom{n}{3}\frac{X_3}{X_1^{n+1}}$$

$$A_{n,n-3} = -\binom{n}{4}\frac{X_4}{X_1^{n+1}} + 10\binom{n+1}{5}\frac{X_2X_3}{X_1^{n+2}} - 15\binom{n+2}{6}\frac{X_2^3}{X_1^{n+3}}$$

$$A_{n,n-4} = -\binom{n}{5}\frac{X_5}{X_1^{n+1}} + 15\binom{n+1}{6}\frac{X_2X_4}{X_1^{n+2}} + 10\binom{n+1}{6}\frac{X_3^2}{X_1^{n+2}}$$
$$- 105\binom{n+2}{7}\frac{X_2^2X_3}{X_1^{n+3}} + 105\binom{n+3}{8}\frac{X_2^4}{X_1^{n+4}}$$

17. Inversion of power series [p. 43]. a) A formal power series $f(x) = a_1x + a_2x^2 + \cdots \in \mathcal{K}[[x]]$ has a compositional inverse, if and only if $a_1 \neq 0$. The inverse is unique and has the form $\overline{f}(x) = b_1x + b_2x^2 + \cdots$, where

$$a_1b_1 = 1$$
$$a_1b_2 + a_2b_1^2 = 0$$
$$a_1b_3 + 2a_2b_1b_2 + a_3b_1^3 = 0$$

$$\vdots$$

See, e. g., [95, Proposition 5.4.1]. It can readily be shown that there exists a unique Laurent polynomial $\Lambda_n^\circ \in \mathcal{K}[X_1^{-1}, X_1, \ldots, X_n]$ such that $b_n = \Lambda_n^\circ(a_1, \ldots, a_n)$ for $n = 1, 2, \ldots$ (Exercise). — We call Λ_n° *ordinary Lagrange inversion polynomial*.

b) E. T. Whittaker [104] has established the following determinantal expression for Λ_n°:

$$\Lambda_n^\circ = \frac{(-1)^{n-1}}{n! X_1^{2n-1}} \begin{vmatrix} nX_2 & X_1 & 0 & 0 & \cdots \\ 2nX_3 & (n+1)X_2 & 2X_1 & 0 & \cdots \\ 3nX_4 & (2n+1)X_3 & (n+2)X_2 & 3X_1 & \cdots \\ 4nX_5 & (3n+1)X_4 & 2(n+1)X_3 & (n+3)X_2 & \cdots \\ \vdots & \vdots & \vdots & \vdots & \end{vmatrix}$$

Denote by f_n and $\overline{f}_n$ the respective coefficients of the Taylor series of $f(x)$ and $\overline{f}(x)$. Then $f_n = n!a_n$ and $\overline{f}_n = n!b_n$, whence

$$\overline{f}_n = n!\Lambda_n^\circ\left(\frac{f_1}{1!}, \ldots, \frac{f_n}{n!}\right).$$

On the other hand, according to Eq. (7.2), $\overline{f}_n = A_{n,1}(f_1, \ldots, f_n)$ holds. We call $\Lambda_n := A_{n,1}$ *exponential Lagrange inversion polynomial*. Obviously both types of Lagrange inversion polynomials can be expressed by each other; for example:

$$\Lambda_n^\circ = \frac{1}{n!}\Lambda_n(1!X_1, \ldots, n!X_n).$$

c) Using a contour integral, whose residue is just the function $\overline{f}(x)$, Morse and Feshbach [71] have succeeded in representing Λ_n° as a partition polynomial, the sum of which is taken over all partition types $(r_1, r_2, r_3, \ldots) \in \mathbb{P}_{n-1}$ (= the union of all sets $\mathbb{P}(n-1, k)$ with $1 \le k \le n-1$). Note that $s := r_1 + r_2 + \cdots$ is a variable assuming values between 1 and $n-1$, while $r_1 + 2r_2 + 3r_3 + \cdots = n-1$ is constant. The result (equation 4.5.12 on p. 412, ibid.) is given here in the form of the exponential Lagrange inversion polynomial:

$$\Lambda_n = A_{n,1} = \frac{1}{X_1^{2n-1}} \sum_{\mathbb{P}_{n-1}} (-1)^s(n-1+s)! \frac{X_1^{n-1-s} X_2^{r_1} X_3^{r_2} \cdots}{r_1! r_2! \cdots (2!)^{r_1}(3!)^{r_2} \cdots}.$$

d) In Riordan's treatise on *Combinatorial Identities* [78] Section 5.4 is titled ›Polynomials for derivatives of inverse functions‹. In this context, a family of polynomials Z_n is established, which in our terminology can be expressed as

$$Z_n(X_1, \ldots, X_n) = A_{n+1,1}(1, -X_1, \ldots, -X_n).$$

Obviously, Z_n computes the Taylor coefficients $\bar{g}_n$ of the inverse function of $g(x) = x - \sum_{n \geq 2} g_{n-1} \frac{x^n}{n!}$, that is, we have $\bar{g}_n = Z_n(g_1, \ldots, g_n)$.

The polynomials Z_1 to Z_8 are listed in Table 5.2 (ibid., p. 181).

Equations (13) and (15) on p. 179 essentially correspond to Comtet's famous Theorem E [22, p. 150-151]; they yield the identity

$$Z_n(X_1, \ldots, X_n) = \sum_{k=1}^{n} B_{n+k,k}(0, X_1, \ldots, X_n).$$

18. Stirling polynomials and Stirling numbers [p. 47]. In Section 8, I somewhat carelessly called the Stirling polynomials ›a kind of strong generalization‹ of the Stirling numbers. Ultimately, however, that statement does not really agree very well with the facts at hand. For example, the polynomial

$$A_{6,2} = \frac{945 X_2^4}{X_1^{10}} - \frac{840 X_3 X_2^2}{X_1^9} + \frac{105 X_4 X_2}{X_1^8} + \frac{70 X_3^2}{X_1^8} - \frac{6 X_5}{X_1^7}$$

obviously conserves much more information than the number $s_1(6, 2) = 274$ that comes about as the sum of its coefficients (representing numbers of certain partitions of a given type).

From a logical point of view, however, it is of course correct to say that $s_1(6, 2)$ is obtained from $A_{6,2}$ through specialization ($X_1 = \cdots = X_5 = 1$).

II

INVERSE RELATIONS AND RECIPROCITY LAWS

1 Introduction

The history of the Bell polynomials essentially originates with the problem of expanding a composite function $f \circ \varphi$ into a Taylor series. Faà di Bruno's famous formula [23, 42] for the higher derivatives of $f \circ \varphi$ provides a solution that takes on an elegant form using the (partial) Bell polynomials $B_{n,k}$:

$$(f \circ \varphi)^{(n)} = \sum_{k=0}^{n} (f^{(k)} \circ \varphi) \cdot B_{n,k}(\varphi', \varphi'', \ldots, \varphi^{(n-k+1)}). \tag{1.1}$$

A wider scope of interesting properties and applications has been opened up, among other things, through the investigations of Bell [6], Riordan [77, 78], and Comtet [22]. And even in the past two decades, research on Bell polynomials has still received considerable attention, as evidenced by a wealth of relevant literature, e. g. [1, 10, 19, 27, 28, 29, 66, 68, 84, 102].

In this chapter a unifying framework is developed for dealing with Bell polynomials and related extensions on an exclusively algebraic basis, which makes it easier to bring general concepts into play and to avoid *ad-hoc* calculations as far as possible.

Above all, the polynomial substitution introduced in this context proves to be an efficient tool. It can already be roughly read from the Faà di Bruno formula that there is a correspondence between functions and polynomials. This basic observation is reflected in rules, according to which the composition of functions corresponds to a specific form of substituting a family of polynomials into a multivariable polynomial.

Another key concept that is directly related to this is inversion. The following variant was a starting point for the author's previous study [84]. If we replace φ in (1.1) by its compositional inverse $\overline{\varphi}$, the question naturally arises whether $B_{n,k}(\overline{\varphi}', \overline{\varphi}'', \ldots, \overline{\varphi}^{(n-k+1)})$ can be represented by a polynomial expression in $\varphi', \varphi'', \ldots, \varphi^{(n-k+1)}$. This problem actually has a solution in the form of a Laurent polynomial $A_{n,k}$ such that

$$B_{n,k}(\overline{\varphi}', \overline{\varphi}'', \ldots, \overline{\varphi}^{(n-k+1)}) = A_{n,k}(\varphi', \varphi'', \ldots, \varphi^{(n-k+1)}) \circ \overline{\varphi}. \tag{1.2}$$

Todorov [99] presumably was the first to establish at least a semi-explicit (determinantal) expression for $A_{n,k}$. A few years earlier Comtet [22] had represented the subfamily $A_{n,1}$ by means of $B_{n-1+k,k}(0, X_2, \ldots, X_n)$, $0 \le k \le n-1$ (see Eq. (7.2) below). In [84], the Comtet formula was extended to the entire family $A_{n,k}$, which eventually also enabled an explicit representation in the form of a partition polynomial (cf. Theorem 2.3 and Eq. (2.16) below). A number of fundamental properties have also been proven, among which the orthogonality relation deserves special emphasis:

$$\sum_{j=k}^{n} A_{n,j} B_{j,k} = \delta_{nk} \qquad (1 \le k \le n). \tag{1.3}$$

Eq. (1.3) is the perfect analogue of the corresponding relation between the Stirling numbers, for in addition to the well-known fact that $B_{n,k}(1, \ldots, 1)$ is equal to the Stirling number of the second kind $s_2(n, k)$, it indeed turns out that $A_{n,k}(1, \ldots, 1)$ is equal to the *signed* Stirling number of the first kind $s_1(n, k)$. In other interesting cases, too, identities that are satisfied by the Stirling numbers can be raised to the level of the corresponding polynomials. To name just a few examples, we mention a theorem by Khelifa and Cherruault [47] (Section 5.4), the introduction of multivariate Lah polynomials (Section 5.5), and the reciprocity theorem $A_{n,k} = (-1)^{n-k} B_{-k,-n}$ (Section 8). The latter shows that the partial Bell polynomials and their orthogonal companions ultimately belong to *one* kind of multivariate *Stirling polynomials*.

The content of the chapter is organized as follows:

Section 2.— In [84] it was sufficient to axiomatically describe the required algebra of functions. However, in order to continue and deepen these investigations it proves advantageous using a suitable standard model instead, here the algebra of formal power series over a field of characteristic zero. Section 2 summarizes the notations used and some of the results required from Chapter I.

Section 3.— For any function φ with Taylor coefficients $\varphi_0, \varphi_1, \varphi_2, \ldots$, a higher-order derivative operator $\Omega_n(\cdot \mid \varphi)$ is introduced that assigns to every function term f (whether containing φ or not) a polynomial $\Omega_n(f \mid \varphi)$ by replacing in $f^{(n)}(0)$ each occurence of φ_j with X_j ($j = 0, 1, 2, \ldots$). Based on rules for evaluating $\Omega_n(f \mid \varphi)$ for composite function terms, we obtain several instances and classes of polynomials that are closely related to the $B_{n,k}$ and will play a crucial role in subsequent sections.

Section 4.— The effect is studied the composition of functions has on polynomials, which depend in a specific way on those functions. Two composition rules will be proved in this context. The first one is the polynomial counterpart of a functional identity $h = f \circ g$. The second rule reformulates a theorem of Jabotinsky and is given here a new proof. It appears as an indispensable tool in many of the proofs to follow.

Section 5.— This section deals with the class of B-representable polynomials that can be written in the form $Q_{n,k} = B_{n,k}(H_1, \ldots, H_{n-k+1})$ for all $n \geq k \geq 1$. Section 5.1 contains some criteria (necessary, sufficient) and the simple fact that a regular B-representable family of polynomials $Q_{n,k}$ has a unique orthogonal companion $Q_{n,k}^{\perp}$. In Section 5.2 it is shown that identities which are valid for Stirling numbers — such as the Stirling inversion or the Schlömilch-Schläfli formulas — can be extended to regular B-representable polynomials. In the remaining subsections, special B-representable polynomial families and their orthogonal companions are examined.

Section 6.— As is well-known there is a close connection between the complete Bell polynomials and binomial sequences. Therefore it appears reasonable to apply to this area the results on Bell polynomials obtained up to then. Most identities known from the literature are direct consequences from our previously established statements. Special interest deserves the binomial sequence related to trees that has been studied by Knuth and Pittel [52] and is given here a new explicit representation. Contributions are also made to the theorem of Mullin and Rota. We give a new proof, supplemented by a 'both-or-none statement', which generalizes a lemma of Yang [105].

Section 7.— Lagrange's classical formula for the inversion of a power series has been generalized in numerous forms (see, e. g., Gessel [30, Theorem 2.1.1]). In this section we are, loosely speaking, concerned with the problem of constructing multivariate polynomials, which convert a sequence of constants that characterizes a given function f into the corresponding sequence of constants that characterizes the inverse function $\overline{f}$. Theorem 7.6 explicitly describes the intricate form of these 'generalized Lagrange inversion polynomials'. Some special cases are discussed in detail, in particular a corollary (= Theorem 7.1) that proves to be equivalent to Comtet's Theorem F [22, p. 151].

Section 8.— This final section is about certain laws of reciprocity. The formulation of such laws requires that the domain of the indices of the polynomials involved can be extended to the integers. In order to achieve this, a new and straightforward procedure is proposed (based on the results from Section 3). The above-mentioned reciprocity law can thus be generalized to any regular B-representable families of polynomials $Q_{n,k}$. The main result

(Theorem 8.1) then states that $Q_{n,k}^{\perp} = (-1)^{n-k}Q_{-k,-n}$ for all $n, k \in \mathbb{Z}$. Finally, two reciprocities for the potential polynomials are derived. The first implies the well-known Schur-Jabotinsky theorem (cited in [30]); the second extends Comtet's Theorem C [22, p. 142] and is shown to be essentially a general version of a binomial transformation attributed to Melzak.

2 Basic notions and preliminaries

2.1 Algebra of functions

Throughout this chapter $\mathcal{K}$ is supposed to be a fixed commutative field of characteristic zero (so that $\mathbb{Q} \subseteq \mathcal{K}$). We denote by $\mathcal{F} := \mathcal{K}[[x]]$ the algebra of formal power series in x with coefficients in $\mathcal{K}$. As customary, addition and scalar multiplication is defined coordinatewise, and product is defined by the Cauchy convolution. The elements $f \in \mathcal{F}$ will be called *functions*. Besides their variable-free notation, we equally write $f(x)$ if the 'argument' x is to be referred in any way. So, the coefficient of x^n in $f(x) = \sum_{n \geq 0} c_n x^n$ is written $c_n = [x^n]f(x)$. We denote by $\mathcal{F}_0$ the set of all $f \in \mathcal{F}$ whose leading coefficient, $[x^0]f(x)$, is zero; furthermore we write $\mathcal{F}_1$ for the complement of $\mathcal{F}_0$ in $\mathcal{F}$. Then, $\mathcal{F}_1$ is the subring of the units of $\mathcal{F}$, that is, its elements are precisely those functions $f \in \mathcal{F}$ having a multiplicative inverse, from now on denoted by f^{-1} or by $1/f$ and called *reciprocal of* f. Another subalgebra of $\mathcal{F}$ is the ring of polynomials, $\mathcal{P} := \mathcal{K}[x]$. When dealing with multivariate Stirling polynomials, it proves appropriate to include certain elements of $\mathcal{K}(x)$, in particular $1/x$. Therefore, we also admit as functions Laurent polynomials, i.e., elements of $\widehat{\mathcal{P}} := \mathcal{K}[x^{-1}, x]$, and only once (in Section 8, Remark 8.3) even formal Laurent series over $\mathcal{K}$.

Next we introduce, as a third binary operation, the *composition* $\circ$ of functions by the following rule of substitution:

$$(f \circ g)(x) := f(g(x)) = \sum_{n \geq 0} ([x^n]f(x))g(x)^n. \tag{2.1}$$

This, however, defines but a partial operation. For almost all[1] of our purposes, two cases will suffice, in which (2.1) makes good sense:

---

[1] The few times (cf. Remark 4.2) that we shall have to treat Laurent polynomials as composite functions $f \circ g$, we will limit ourselves to the case $f \in \mathcal{P}, g \in \widehat{\mathcal{P}}$.

$$f \in \mathcal{F} \text{ and } g \in \mathcal{F}_0 \quad \text{(called 0-\textit{case})},$$
$$f \in \widehat{\mathcal{P}} \text{ and } g \in \mathcal{F}_1 \quad \text{(called 1-\textit{case})}.$$

In either case we get a well-defined composite function $f \circ g$. The identity element is $\iota = \iota(x) := x$, satisfying $f \circ \iota = \iota \circ f = f$. Moreover, we have $f \circ (g \circ h) = (f \circ g) \circ h$, $(f + g) \circ h = (f \circ h) + (g \circ h)$, and $(f \cdot g) \circ h = (f \circ h) \cdot (g \circ h)$, whenever the terms involved are meaningful. By $\mathcal{G}$ we denote the set of $g \in \mathcal{F}_0$, for which a compositional inverse $f \in \mathcal{F}_0$ exists satisfying $f \circ g = g \circ f = \iota$. We write $\bar{g}$ for the (unique) inverse of g. The set of invertible functions in $\mathcal{F}$ forms a non-abelian group with composition. Recall that $\mathcal{G} = \{g \in \mathcal{F}_0 \mid [x]g(x) \neq 0\}$.

Remark 2.1. We note the following simple, but useful statement: *The product of two functions in $\mathcal{F}$ is invertible, if and only if either is invertible while the other is a unit.* From this follows immediately

$$f \in \mathcal{G} \iff \iota^{-1} \cdot f \in \mathcal{F}_1. \tag{2.2}$$

Among the few special functions we shall be concerned with further on, consider the exponential function in $\mathcal{F}$ defined as usual by $\exp(x) := 1 + x + x^2/2! + x^3/3! + \cdots$ (occasionally written in traditional notation e^x). Since $\exp$ belongs to $\mathcal{F}_1$, it has no compositional inverse, and $\log$ cannot be represented in $\mathcal{F}$. Alternatively, we define $\varepsilon := \exp - 1$, which has Mercator's series for $\log(1 + x)$ as its inverse in $\mathcal{G}$, namely

$$\lambda(x) := \bar{\varepsilon}(x) = \sum_{n \geq 1} (-1)^n \frac{x^n}{n}.$$

Finally, we need our algebra of functions to be endowed with a *derivation*, that is, an additive operator D satisfying the Leibniz product rule $D(fg) = fD(g) + D(f)g$. A derivation is said to be *normalized* when it takes x to 1. A normalized derivation D on $\mathcal{P}$ agrees with the ordinary derivation known from calculus: $D(a_0 + a_1 x + a_2 x^2 + \cdots + a_n x^n) = a_1 + 2a_2 x + \cdots + na_n x^{n-1}$. Recall now that D can be extended in a unique way to a derivation on the rational functions (quotient field of $\mathcal{P}$) as well as to a derivation on $\mathcal{F}$ (see, e.g., [53, p. 31], and for some more details [7]). We still denote these extensions by D (only some few times writing f' instead of $D(f)$). As a consequence, the chain rule $D(f \circ g) = (D(f) \circ g) \cdot D(g)$ is satisfied in both the 0-case and the 1-case as well.

Iterating D leads, in the usual way, to derivatives of higher order $D^n(f)$, $n \geq 0$. For example, applied to ε, λ, and ι^{-1} one has for all $n \geq 1$:

$$D^n(\varepsilon) = D^n(\exp) = \exp = 1 + \varepsilon, \tag{2.3}$$

$$D^n(\lambda) = (-1)^{n-1}(n-1)!(1+\iota)^{-n}, \tag{2.4}$$

$$D^n(\iota^{-1}) = (-1)^n n! \iota^{-(n+1)}. \tag{2.5}$$

Here, as in all similar cases, the term $(1 + \iota)^{-n}$ in (2.4) is to be understood as the nth power of $(1+\iota)^{-1}$ which, of course, is the function (geometric series) $1 - \iota + \iota^2 - \iota^3 + \cdots \in \mathcal{F}_1$. Generally, every $f \in \mathcal{F}$ can be written in form of a Taylor series $f(x) = \sum_{n \geq 0} D^n(f)(0)x^n/n!$.

We conclude this subsection with some remarks concerning the operator θD, which is a derivation whenever $\theta \in \mathcal{F}$. Comtet [21] called θD *Lie derivation* (with respect to the function θ) and defined it by $\theta D(f)(x) := \theta(x)f'(x)$ (see also [69]). Since long the special case $\theta = \iota$ as well as the nice expansions resulting from repeatedly applying xD to a function had been studied in the literature (see, e. g., [13, 44, 59, 101]). However, leaving θ unspecified, one obtains far less satisfactory results (an issue we shall come back to in Section 5.6). As is shown in [84] (= Chapter I), the situation takes a happy turn when, following Todorov [99], one chooses θ to be of the form $D(\varphi)^{-1}$. Therefore, given any function φ with $D(\varphi)(0) \neq 0$, we define D_φ by

$$D_\varphi(f) := D(\varphi)^{-1}D(f). \tag{2.6}$$

The operator D_φ may be called *derivation with respect to* φ, because it acts on a function as if φ were its independent variable, for instance: $D_\varphi(1+3\varphi-\varphi^5) = 3 - 5\varphi^4$, or $D_\varphi(e^\varphi) = e^\varphi$.

Remark 2.2. A peculiar detail: Just the classical operator xD cannot be represented as a D_φ, since $\varphi = \log$ is not a function in our sense. Alternatively, one could use $D_{\log}$ as an abbreviation for $D_\lambda - D$ thus achieving the analogous effect a genuine operator $D_{\log}$ would have on, say, meromorphic functions.

For an arbitrary $\varphi \in \mathcal{G}$ and $n \geq 0$ it can easily be shown [84, Proposition 2.2: *Pourchet's formula*] that

$$D_\varphi^n(f) = D^n(f \circ \overline{\varphi}) \circ \varphi. \tag{2.7}$$

Specializing $f = \iota$ in (2.7), we get the nth Taylor coefficient of the inverse function $\overline{\varphi}$ by just interchanging ι and φ in $D_\iota^n(\varphi)(0)$ ($=$ the nth Taylor coefficient of φ), i. e., we have $D^n(\overline{\varphi})(0) = D_\varphi^n(\iota)(0)$. Compared to the classical term $D^{n-1}((\iota/\varphi)^n)(0)$, which Lagrange proposed to obtain the inverse of a power series, the iterative expression $D_\varphi^n(\iota)(0)$ proves simpler and, in most cases, requires significantly less computational amount.

The next subsection will exhibit the important part the operator D_φ plays in setting up a suitable framework for dealing with multivariate Bell polynomials and their related extensions.

2.2 Multivariate polynomials

We consider polynomials over the field $\mathcal{K}$ as well as some kinds of special Laurent polynomials from $\mathcal{K}[X_0^{-1}, X_1^{-1}, X_0, X_1, \ldots, X_n]$ which, for brevity, will be referred as polynomials, too. The partial derivation on $\mathcal{K}[X_1, \ldots, X_n]$ with respect to X_j can be regarded as the (unique) normalized derivation on $\mathcal{K}'[X_j]$ where $\mathcal{K}' = \mathcal{K}[X_1, \ldots, X_{j-1}, X_{j+1}, \ldots, X_n]$; it extends in a unique way to the rational functions $\mathcal{K}(X_1, \ldots, X_n)$ and is denoted by $\frac{\partial}{\partial X_j}$.

Given a polynomial P in $X_1, X_2, \ldots$ and any sequence of polynomials (Q_n), we denote by $P \circ Q_\sharp$ the result obtained by replacing in P each indeterminate X_j by Q_j (the $\sharp$-sign marks the indexed place that corresponds to the indeterminate's index). Besides $P \circ Q_\sharp$, we also use the traditional notation $P(Q_1, Q_2, \ldots)$. When the sequence $Q_1, Q_2, \ldots$ is constant, the substitution will be called *unification*, and $\sharp$ be dropped. In the (default) case $Q_1 = Q_2 = \cdots = 1$ we plainly write $P \circ 1$, which gives the sum of the coefficients of P.

A reasonable sense can also be attached to expressions of the form $Q_\sharp \circ R_\sharp$. Suppose $Q_n \in \mathcal{K}[X_1, \ldots, X_{q(n)}]$, with an integer $q(n) \geq 1$. We then take $Q_\sharp \circ R_\sharp$ as an abbreviation for $Q_\sharp(R_1, \ldots, R_{q(\sharp)})$. The composition of polynomials obeys the *associative law*:

$$(P \circ Q_\sharp) \circ R_\sharp = P \circ (Q_\sharp \circ R_\sharp). \tag{2.8}$$

Given any function φ, consider the mapping $P \mapsto P^\varphi := P \circ D^\sharp(\varphi)$, which assigns to each polynomial P the function obtained from P by replacing X_j by $D^j(\varphi)$ for each j. The following substitution rule is obvious:

$$P(Q_1, \ldots, Q_n)^\varphi = P(Q_1^\varphi, \ldots, Q_n^\varphi). \tag{2.9}$$

Remark 2.3. Let P, Q be polynomials in $X_1, X_2, \ldots$ satisfying $P^\varphi(0) = Q^\varphi(0)$ for every $\varphi \in \mathcal{F}$. We then have $P = Q$. — Since the constants in any given sequence $c_1, c_2, \ldots \in \mathcal{K}$ may be regarded as Taylor coefficients

$c_j = D^j(\varphi)(0)$ of some function φ, one merely has to recall that distinct polynomials over an infinite integral domain cannot give rise to the same polynomial function.

In the sequel double-indexed families of polynomials play a major role. We use the notation $(U_{n,k})$ with $n, k \geq 0$ to mean the infinite family, and moreover $(U_{i,j})_{0 \leq i,j \leq n}$ to denote its initial part in form of a quadratic matrix. The family (and the matrix as well) is called *(lower) triangular* if $U_{n,k} = 0$ for $n < k$. We say that polynomial families $(U_{n,k})$, $(V_{n,k})$ are *orthogonal companions* (of each other) when they satisfy the *orthogonality relation*

$$\sum_{j=0}^{n} U_{n,j} V_{j,k} = \delta_{nk} \tag{2.10}$$

for all $n, k \geq 0$ (δ_{nk} Kronecker's symbol). In this case we equally write $U_{n,k} = V_{n,k}^{\perp}$ or $V_{n,k} = U_{n,k}^{\perp}$. The families involved will be called *regular* inasmuch they necessarily have non-singular matrices.

Next, we summarize without proofs some relevant material from the author's paper on multivariate Stirling polynomials the following sections are based upon. For details we refer the reader to Chapter I.

Unless otherwise stated we assume φ to be an arbitrary function from $\mathcal{G}$, and $i, j, k, n, \ldots$ to be non-negative integers.

Proposition 2.1. *There exist polynomials $A_{n,k} \in \mathcal{K}[X_1^{-1}, X_2, \ldots, X_{n-k+1}]$ such that*

$$D_{\varphi}^n = \sum_{k=0}^{n} A_{n,k}^{\varphi} \cdot D^k.$$

The family $(A_{n,k})$ is triangular, regular, and uniquely determined by the differential recurrence

$$A_{n+1,k} = \frac{1}{X_1} \left(A_{n,k-1} + \sum_{j=1}^{n-k+1} X_{j+1} \frac{\partial A_{n,k}}{\partial X_j} \right), \quad A_{n,0} = \delta_{n0}.$$

The expansion of D_{φ}^n into a linear combination of the $D^0, D^1, \ldots, D^n$ can also be done in the reverse direction.

Proposition 2.2. *There exist polynomials $B_{n,k} \in \mathcal{K}[X_1, X_2, \ldots, X_{n-k+1}]$ such that*

$$D^n = \sum_{k=0}^{n} B_{n,k}^{\varphi} \cdot D_{\varphi}^k.$$

The family $(B_{n,k})$ is triangular, regular, and uniquely determined by the differential recurrence

$$B_{n+1,k} = X_1 \left(B_{n,k-1} + \sum_{j=1}^{n-k+1} X_{j+1} \frac{\partial B_{n,k}}{\partial X_j} \right), \quad B_{n,0} = \delta_{n0}.$$

Both polynomial families are closely connected, as becomes evident by the identity

$$A_{n,k}^{\varphi} = B_{n,k}^{\overline{\varphi}} \circ \varphi. \tag{2.11}$$

Taking $\overline{\varphi}$ for φ, it can be equivalently expressed in the form $B_{n,k}^{\varphi} = A_{n,k}^{\overline{\varphi}} \circ \varphi$.

To start with Proposition 2.2: $(B_{n,k})$ are the (*partial*) *Bell polynomials* Riordan [77] named in honor of E. T. Bell, whose paper [6] is an extensive study of what Bell himself called 'exponential polynomials', obviously motivated by the eigenvalue identity $D^n(e^{\varphi}) = (B_{n,0} + B_{n,1} + \cdots + B_{n,n})^{\varphi} \cdot e^{\varphi}$. The latter results as the special case $f = \exp$ of the famed *Faà di Bruno* (FdB) *formula*

$$D^n(f \circ \varphi) = \sum_{k=0}^{n} (D^k(f) \circ \varphi) \cdot B_{n,k}^{\varphi}, \tag{2.12}$$

which appears as a byproduct in the proof of Proposition 2.2. Also wellknown is the following 'diophantine' representation

$$B_{n,k} = \sum_{\mathbb{P}(n,k)} \frac{n!}{r_1! r_2! \cdots (1!)^{r_1} (2!)^{r_2} \cdots} X_1^{r_1} X_2^{r_2} \cdots X_{n-k+1}^{r_{n-k+1}}, \tag{2.13}$$

the sum to be taken over all elements $(r_1, \ldots, r_{n-k+1})$ of the set $\mathbb{P}(n,k)$ of (n,k)-partition types, i. e., sequences of integers $r_1, r_2, r_3, \ldots \geq 0$ such that $r_1 + r_2 + r_3 + \cdots = k$ and $r_1 + 2r_2 + 3r_3 + \cdots = n$. It follows that $B_{n,k}$ is homogeneous of degree k and isobaric of degree n.

Let us now turn to Proposition 2.1 and to the most fundamental properties of the family $(A_{n,k})$. The first thing to notice here is the fact that $(A_{n,k})$ is the orthogonal companion of the Bell polynomials: $B_{n,k}^{\perp} = A_{n,k}$.

Remark 2.4. Let $\varphi(x) = c_1 x + c_2 x^2/2! + c_3 x^3/3! + \cdots$, $c_1 \neq 0$. Since $B_{n,1} = X_n$, we obtain from Eq. (2.11) $A_{n,1}^{\varphi}(0) = (A_{n,1}^{\varphi} \circ \overline{\varphi})(0) = B_{n,1}^{\overline{\varphi}}(0) = D^n(\overline{\varphi})(0)$, that is, $A_{n,1}(c_1, \ldots, c_n)$ is the nth Taylor coefficient of the inverse function $\overline{\varphi}$.

Thus we are led to the following compositional identities, which can be viewed as polynomial counterparts of Eq. (2.11):

$$A_{n,k} = B_{n,k} \circ A_{\sharp,1} = B_{n,k}(A_{1,1}, \ldots, A_{n-k+1,1}) \qquad (2.14)$$

$$B_{n,k} = A_{n,k} \circ A_{\sharp,1} = A_{n,k}(A_{1,1}, \ldots, A_{n-k+1,1}). \qquad (2.15)$$

Set $\widetilde{B}_{n,k} := B_{n,k}(0, X_2, \ldots, X_{n-k+1})$, called *associate* (partial) *Bell polynomials* (the coefficients of which count only partitions with no singleton blocks). Then the main result[2] of [84] is as follows:

Theorem 2.3. *Let* k, n *be any integers with* $1 \le k \le n$. *We have*

$$A_{n,k} = \sum_{j=k-1}^{n-1} (-1)^{n-1-j} \binom{2n-2-j}{k-1} X_1^{j-2n+1} \widetilde{B}_{2n-1-k-j,n-1-j}.$$

This identity strongly generalizes Comtet's famous formula for the coefficients of an inverted power series [22, Theorem E, p. 150/151], here obtained by taking $k = 1$. Theorem 2.3 can be used to derive some general polynomial identities of Schlömilch-Schläfli type (see Section 5.2) as well as the following explicit representation, which corresponds to that of $B_{n,k}$ in Eq. (2.13):

$$A_{n,k} = X_1^{-(2n-1)} \sum_{\mathbb{P}(2n-1-k,n-1)} \frac{(-1)^{n-1-r_1}(2n-2-r_1)!}{(k-1)!r_2!r_3!\cdots(2!)^{r_2}(3!)^{r_3}\cdots} X_1^{r_1} X_2^{r_2} \cdots$$

$$(2.16)$$

As a consequence, $A_{n,k}$ is homogeneous of degree $n-1$ and isobaric of degree $2n-1-k$. — Unification with both $A_{n,k}$ and $B_{n,k}$ yields

$$A_{n,k} \circ 1 = s_1(n,k) \text{ (signed Stirling numbers of the first kind),} \qquad (2.17)$$

$$B_{n,k} \circ 1 = s_2(n,k) \text{ (Stirling numbers of the second kind).} \qquad (2.18)$$

This may justify $A_{n,k}$ and $B_{n,k}$ being called *multivariate Stirling polynomials* (MSP) *of the first and second kind*, respectively. While Eq. (2.18) has been well-known since long, (2.17) does add a new facet to what is already known about how the *unsigned* Stirling numbers $c(n,k) := |s_1(n,k)|$ could be obtained by unification from certain polynomials (see Sections 5.2 and 5.5).

We finally state a property of the Bell polynomials that will be needed later.

---

[2] In [84] this statement was proved by an inductive argument. It has turned out later that an independent proof can be provided with the help of the reciprocity law (see Eq. (8.4) and Remark 8.2 below).

Lemma 2.4 (Identity Lemma). *Let $\varphi, \psi \in \mathcal{G}$ be any invertible functions such that $B_{n,k}^{\varphi}(0) = B_{n,k}^{\psi}(0)$ holds for all n, k with $1 \leq k \leq n$. Then $\varphi = \psi$.*

Proof. Since this statement does not occur in [84], we will at least sketch a proof (by induction). Let $\varphi_1, \varphi_2, \varphi_3, \ldots$ and $\psi_1, \psi_2, \psi_3, \ldots$ be the Taylor coefficients of φ and ψ, respectively, and let $n \geq 1$ be an arbitrary integer. It suffices to show that $B_{n,n-k}(\varphi_1, \ldots, \varphi_{k+1}) = B_{n,n-k}(\psi_1, \ldots, \psi_{k+1})$ implies $\varphi_j = \psi_j$ $(j = 1, \ldots, k+1)$ for $k = 0, 1, \ldots, n-1$. — For $k = 0$ we have $\varphi_1^n = B_{n,n}^{\varphi}(0) = B_{n,n}^{\psi}(0) = \psi_1^n$. Taking into account that $\varphi_1, \psi_1 \neq 0$ and that the equation must hold for an arbitrary n, this implies $\varphi_1 = \psi_1$. Now the induction step $k \to k+1$ can be carried out using the formula

$$B_{n,n-k} = \sum_{i=1}^{k} X_{i+1} \cdot \sum_{j=0}^{n-k-1} \binom{n-j-1}{i} X_1^j B_{n-j-i-1,n-j-k-1}. \quad (2.19)$$

This may be left to the reader. One derives[3] Eq. (2.19) from the recurrence in Proposition 2.2 with the help of the identity

$$\frac{\partial B_{n,k}}{\partial X_j} = \binom{n}{j} B_{n-j,k-1} \quad (1 \leq j \leq n - k + 1). \quad (2.20)$$

See [84, Corollary 4.4] (while [6, Equation (5.1), p. 266] is a version of Eq. (2.20) for the complete Bell polynomials B_n). — It should be noticed here that the Taylor coefficients with index ≥ 2 occur only in linear form. We illustrate this for the induction step in the case $k = 1$. Assuming $B_{n,n-1}(\varphi_1, \varphi_2) = B_{n,n-1}(\psi_1, \psi_2)$ we would have by Eq. (2.19) $\binom{n}{2}\varphi_1^{n-2}\varphi_2 = \binom{n}{2}\psi_1^{n-2}\psi_2$, which yields $\varphi_2 = \psi_2$. $\Diamond$

3 Polynomials from Taylor coefficients

Throughout this section $\varphi \in \mathcal{F}$ takes on the role of a fixed unspecified placeholder for an arbitrary function. The mapping $P \mapsto P^{\varphi}$ considered above that makes every polynomial into a function, could easily be reversed by replacing each $D^j(\varphi)$ in P^{φ} by X_j. This can, more generally, be done in a slightly modified way for any function terms, which designate a function

---

[3] For an outline see Chapter I, Notes and supplements no. 7.

in $\mathcal{F}$. To achieve this, we will introduce a higher-order derivative opera-
tor $\Omega_n(\cdot \,|\, \varphi)$ assigning to each such term f a polynomial $\Omega_n(f \,|\, \varphi)$ with the
property

$$\Omega_n(f \,|\, \varphi)^\varphi(0) = D^n(f)(0). \tag{3.1}$$

Since there seems to be no real risk of misunderstandings, we will not distinguish in our notation between functions and function terms. It may suffice
here, on a more informal basis and without switching explicitly to the meta-
level of logical syntax, to remind that function terms are built up inductively
from simpler ones. Thus, saying that φ occurs in f (or: f contains φ) is to
be understood by the usual method of recursion on terms in the sense that
φ occurs in φ and, if φ occurs in f or in g, then φ also occurs in composite
terms like $f + g$, $f \cdot g$, f^{-1}, $f \circ g$, and $\overline{f}$.

Definition 3.1. Let f be any function term. Then $\Omega_n(f \,|\, \varphi)$ is obtained
by replacing for each $j \geq 0$ the occurences of $D^j(\varphi)(0)$ in $D^n(f)(0)$ with
the indeterminate X_j. (Thus $\Omega_n(f \,|\, \varphi)$ results as a polynomial over $\mathcal{K}$ that
satisfies Eq. (3.1)).

Some simple cases are immediate:

$$\Omega_n(f \,|\, \varphi) = D^n(f)(0), \text{ whenever } \varphi \text{ does not occur in } f. \tag{3.2}$$

For example, one has $\Omega_n(c \,|\, \varphi) = \delta_{n0} \cdot c \ (c \in \mathcal{K})$, $\Omega_n(\iota^k \,|\, \varphi) = \delta_{nk} \cdot k!$,
$\Omega_n(\varepsilon \,|\, \varphi) = 1 - \delta_{n0}$, $\Omega_n(\lambda \,|\, \varphi) = s_1(n, 1)$. It is also obvious that

$$\Omega_n(\varphi \,|\, \varphi) = X_n. \tag{3.3}$$

Since D^n is additive and satisfies the Leibniz rule for higher-order derivatives,
we have

$$\Omega_n(f + g \,|\, \varphi) = \Omega_n(f \,|\, \varphi) + \Omega_n(g \,|\, \varphi), \tag{3.4}$$

$$\Omega_n(f \cdot g \,|\, \varphi) = \sum_{k=0}^{n} \binom{n}{k} \Omega_{n-k}(f \,|\, \varphi) \Omega_k(g \,|\, \varphi). \tag{3.5}$$

In particular, $\Omega_n(c \cdot f \,|\, \varphi) = c \cdot \Omega_n(f \,|\, \varphi)$ holds for every $c \in \mathcal{K}$. Thus $\Omega_n(\cdot \,|\, \varphi)$
operates $\mathcal{K}$-linearly on $\mathcal{F}$.

For the composite term $f \circ g$ we have to treat the 0-case separately from
the 1-case. In the former case we have $f \in \mathcal{F}$ and $g \in \mathcal{F}_0$. According to the

FdB formula (2.12) and observing that $g(0) = 0$, we obtain

$$\Omega_n(f \circ g \mid \varphi) = \sum_{k=0}^{n} \Omega_k(f \mid \varphi) \cdot (B_{n,k} \circ \Omega_\sharp(g \mid \varphi)). \qquad (3.6)$$

In the 1-case we assume g to be a function in $\mathcal{F}_1$, that is, we have $g(0) \neq 0$ and $f \in \widehat{\mathcal{P}}$. Consequently, f cannot contain φ and Eq. (3.6) has to be modified as follows:

$$\Omega_n(f \circ g \mid \varphi) = \sum_{k=0}^{n} D^k(f)(\Omega_0(g \mid \varphi)) \cdot (B_{n,k} \circ \Omega_\sharp(g \mid \varphi)). \qquad (3.7)$$

Remark 3.1. As mentioned above, Definition 3.1 could be given a logically more rigorous form as an induction on terms; then, equations (3.2) to (3.7) would become part of the definition. If one wants also the operator D to be involved in the formation of function terms, then the plausible equation $\Omega_n(D(f) \mid \varphi) = \Omega_{n+1}(f \mid \varphi)$ must be added.

Next we turn to the unary operations taking f to f^{-1} (reciprocation) and to $\overline{f}$ (inversion). Either cases turn out to be derivable from the properties of Ω_n so far established, inasmuch both f^{-1} and $\overline{f}$ can to a sufficient degree be explicitly expressed as functions in $\mathcal{F}$.

Reciprocation. — For any $f \in \mathcal{F}_1$ we have $f(0) \neq 0$, hence

$$\Omega_n(f^{-1} \mid \varphi) = \Omega_n(\iota^{-1} \circ f \mid \varphi)$$

$$= \sum_{k=0}^{n} D^k(\iota^{-1})(\Omega_0(f \mid \varphi)) \cdot (B_{n,k} \circ \Omega_\sharp(f \mid \varphi)) \qquad (\iota^{-1} \in \widehat{\mathcal{P}}, \text{Eq. (3.7)})$$

$$= \sum_{k=0}^{n} (-1)^k k! \, \Omega_0(f \mid \varphi)^{-(k+1)} B_{n,k}(\Omega_1(f \mid \varphi), \ldots, \Omega_{n-k+1}(f \mid \varphi)).$$

This gives rise to

Definition 3.2. Let $\widehat{R}_n \in \mathcal{K}[X_0^{-1}, X_1, \ldots, X_n]$, $n \geq 0$, be the (Laurent) polynomials

$$\widehat{R}_n := \sum_{k=0}^{n} (-1)^k k! \, X_0^{-(k+1)} B_{n,k},$$

henceforth called *reciprocal polynomials*. We shall also use the special case $R_n := \widehat{R}_n(1, X_1, \ldots, X_n)$.

Together with the foregoing, we thus have obtained

Proposition 3.1. $\Omega_n(f^{-1} \,|\, \varphi) = \widehat{R}_n(\Omega_0(f \,|\, \varphi), \ldots, \Omega_n(f \,|\, \varphi))$ *for all* $f \in \mathcal{F}_1$
and $n \geq 0$.

Remark 3.2. We would have been arrived at the same result by applying
Eq. (3.6) instead of Eq. (3.7) to the reciprocal function written as $f^{-1} = f(0)^{-1}((1 + \iota)^{-1} \circ f(0)^{-1}(f - f(0)))$.

Inversion. — The inverse $\overline{f}$ of a function $f \in \mathcal{G}$ is given by its Taylor coefficients $D^n(\overline{f})(0)$, $n \geq 1$. We will use two options to represent these coefficients somewhat more explicitly. First, according to Remark 2.4 we have
$D^n(\overline{f})(0) = A_{n,1}(D(f)(0), D^2(f)(0), \ldots, D^n(f)(0))$, which immediately
yields

Proposition 3.2. $\Omega_n(\overline{f} \,|\, \varphi) = A_{n,1}(\Omega_1(f \,|\, \varphi), \ldots, \Omega_n(f \,|\, \varphi))$ *for all* $f \in \mathcal{G}$
and $n \geq 1$.

Our second option is to make use of the Lagrange formula $D^n(\overline{f})(0) = D^{n-1}((\iota/f)^n)(0)$. It follows from Eq. (2.2) that $\iota/f = (\iota^{-1} \cdot f)^{-1} \in \mathcal{F}_1$,
whence $\Omega_0(\iota/f \,|\, \varphi) = (\iota/f)(0) \neq 0$. Therefore, when passing over from
$D^n(\overline{f})(0)$ to $\Omega_n(\overline{f} \,|\, \varphi)$ we again have to apply Eq. (3.7):

$$\Omega_n(\overline{f} \,|\, \varphi) = \Omega_{n-1}(\iota^n \circ \tfrac{\iota}{f} \,|\, \varphi)$$

$$= \sum_{k=0}^{n-1} D^k(\iota^n)(\Omega_0(\tfrac{\iota}{f} \,|\, \varphi)) \cdot (B_{n-1,k} \circ \Omega_\sharp(\tfrac{\iota}{f}) \,|\, \varphi)$$

$$= \sum_{k=0}^{n-1} (n)_k \Omega_0(\tfrac{\iota}{f} \,|\, \varphi)^{n-k} B_{n-1,k}(\Omega_1(\tfrac{\iota}{f} \,|\, \varphi), \ldots, \Omega_{n-k}(\tfrac{\iota}{f} \,|\, \varphi)),$$

where $(n)_k$ means the falling power $n(n - 1) \cdots (n - k + 1)$, $k \geq 1$, and
$(n)_0 = 1$. — This shows that $\Omega_n(\overline{f} \,|\, \varphi)$ can be represented by another class of
polynomials.

Proposition 3.3. $\Omega_n(\overline{f} \,|\, \varphi) = \widehat{T}_n(\Omega_0(\tfrac{\iota}{f} \,|\, \varphi), \ldots, \Omega_{n-1}(\tfrac{\iota}{f} \,|\, \varphi))$ *for* $n \geq 1$,
where $\widehat{T}_n \in \mathcal{K}[X_0, \ldots, X_{n-1}]$ *is defined by*

$$\widehat{T}_n := \sum_{k=0}^{n-1} (n)_k X_0^{n-k} B_{n-1,k}$$

and called nth tree polynomial (cf. part (iii) of Remark 3.3). *For later use we set*
$T_n := \widehat{T}_n(1, X_1, \ldots, X_{n-1})$.

Corollary 3.4 (Proposition 3.2 and 3.3). *For every* $n \geq 1$

(i) $A_{n,1} = \widehat{T}_n(\widehat{R}_0(\frac{X_1}{1}), \widehat{R}_1(\frac{X_1}{1}, \frac{X_2}{2}), \ldots, \widehat{R}_{n-1}(\frac{X_1}{1}, \frac{X_2}{2}, \ldots, \frac{X_n}{n}))$

(ii) $A_{n,1}(X_0, 2X_1, 3X_2, \ldots) = \widehat{T}_n(\widehat{R}_0, \widehat{R}_1 \ldots, \widehat{R}_{n-1})$

(iii) $A_{n,1}(1, 2X_1, 3X_2, \ldots) = T_n(R_1, R_2, \ldots, R_{n-1})$

Proof. (i). We assume $f(x) = \sum_{n \geq 1} f_n x^n / n!$ to be an invertible function, in which φ does not occur. Thus we have $\Omega_n(f \mid \varphi) = f_n$ and $(\iota^{-1} \cdot f)(x) = \sum_{n \geq 0} \frac{f_{n+1}}{n+1} \frac{x^n}{n!}$, whence by Proposition 3.1

$$\Omega_n(\tfrac{\iota}{f} \mid \varphi) = \widehat{R}_n(\tfrac{f_1}{1}, \tfrac{f_2}{2}, \ldots, \tfrac{f_{n+1}}{n+1}). \tag{3.8}$$

Let $H_n(X_1, \ldots, X_n)$ temporarily stand for the polynomial on the right-hand side of (i). Then, combining Proposition 3.2 with 3.3 and Eq. (3.8) we get

$$A_{n,1}^f(0) = A_{n,1}(f_1, \ldots, f_n) = H_n(f_1, \ldots, f_n) = H_n^f(0).$$

The argument from Remark 2.3 now yields the assertion (i): $A_{n,1} = H_n$.
(ii) and (iii) are immediate consequences of (i). $\diamondsuit$

Later on, part (i) of Corollary 3.4 will be extended to the entire family $(A_{n,k})$ (see Corollary 3.8 below).

3.1 Faà di Bruno polynomials

Let $f \in \mathcal{F}$ be any function, which does not contain φ. For $n \geq 0$ we define the nth FdB polynomial of f (for short, f-*polynomial*) by

$$\Phi_n(f) := \Omega_n(f \circ \varphi \mid \varphi) = \sum_{k=0}^{n} D^k(f)(0) B_{n,k}. \tag{3.9}$$

The sum on the right-hand side results from Eq. (3.6), where $\varphi \in \mathcal{F}_0$ is assumed. This definition makes also sense in the case $f \in \widehat{\mathcal{P}}$, $\varphi \in \mathcal{F}_1$. It is clear then by Eq. (3.7) that the Taylor coefficient $D^k(f)(0)$ in Eq. (3.9) has to be replaced with the Laurent polynomial $D^k(f)(X_0)$.

This definition is still ambiguous for $f \in \mathcal{P}$, insofar $f \circ \varphi$ can be evaluated with both $\varphi \in \mathcal{F}_0$ and $\varphi \in \mathcal{F}_1$. To distinguish the two cases, we shall indicate the choice $\varphi \in \mathcal{F}_1$ by writing $\widehat{\Phi}_n(f)$ instead of $\Phi_n(f)$, if necessary.

We check at once that Φ_n is a $\mathcal{K}$-linear operator obeying the product rule (3.5). The case of a composite function will be treated in Section 4.

Examples 3.1. Well-known examples of FdB polynomials are

(i): the *exponential (or: complete Bell) polynomials* $B_n := \Phi_n(\exp) = \sum_{k=0}^{n} D^k(\exp)(0)B_{n,k} = \sum_{k=0}^{n} B_{n,k}$, $n \geq 0$. We have $\Phi_n(\varepsilon) = B_n$ for $n \geq 1$, however $\Phi_0(\varepsilon) = 0 \neq B_0 = 1$.

(ii): the *logarithmic polynomials* $L_n := \Phi_n(\lambda) = \sum_{k=1}^{n} D^k(\lambda)(0)B_{n,k}$ $\overset{(2.4)}{=} \sum_{k=1}^{n}(-1)^{k-1}(k-1)!B_{n,k}$ (cf. Comtet [22, p. 140]).

(iii): the *potential polynomials* $\widehat{P}_{n,k} := \widehat{\Phi}_n(\iota^k)$ [22, p. 141]. Expanding $\Omega_n(\iota^k \circ \varphi \,|\, \varphi)$ according to Eq. (3.7) gives

$$\widehat{P}_{n,k} = \sum_{j=0}^{n}(k)_j X_0^{k-j} B_{n,j}. \tag{3.10}$$

The reciprocal polynomials and the tree polynomials form subfamilies of $(\widehat{P}_{n,k})$, since we have $\widehat{R}_n = \widehat{P}_{n,-1}$ and $\widehat{T}_n = \widehat{P}_{n-1,n}$.

(iv): The fact that powers and falling powers are connected by $(x)_k = \sum_{j=0}^{k} s_1(k,j)x^j$ suggests introducing *factorial polynomials* by a definition analogous to that of (iii) above: $\widehat{F}_{n,k} := \widehat{\Phi}_n((\iota)_k)$; additionally set $F_{n,k} := \widehat{F}_{n,k}(1, X_1, X_2, \ldots)$. Since $\widehat{\Phi}_n$ is $\mathcal{K}$-linear, we immediately obtain

$$\widehat{F}_{n,k} = \sum_{j=0}^{k} s_1(k,j)\widehat{P}_{n,j}, \tag{3.11}$$

hence by Stirling inversion (see e. g. [94, Proposition 1.4.1b])

$$\widehat{P}_{n,k} = \sum_{j=0}^{k} s_2(k,j)\widehat{F}_{n,j}. \tag{3.12}$$

In a more explicit form the factorial polynomials can be written

$$\widehat{F}_{n,k} = \sum_{r=0}^{n}\left(\sum_{j=r}^{k} s_1(k,j)(j)_r X_0^{j-r}\right)B_{n,r}.$$

Though perhaps not apparent at first glance, the potential polynomials are indeed closely related to the Bell polynomials; so it is worth examining them in more detail. We start with restricting k to non-negative values (thereby choosing $\varphi \in \mathcal{F}_0$). This leads to a special instance of $\widehat{P}_{n,k}$:

$$\Phi_n(\iota^k) = \sum_{j=1}^{n} \Omega_j(\iota^k \mid \varphi) B_{n,j} = \sum_{j=1}^{n} \delta_{jk} \cdot k! B_{n,j}$$

$$= k! B_{n,k} = \widehat{P}_{n,k}(0, X_1, \ldots, X_n). \tag{3.13}$$

Now consider the function (geometric series) $\gamma := (1 - \iota)^{-1} = 1 + \iota + \iota^2 + \ldots$, which belongs to $\mathcal{F}_1$. For the corresponding FdB polynomials we get $\Phi_n(\gamma) = \sum_{k=0}^{n} k! B_{n,k}$, which may be called *geometric polynomials* (in view of the fact that their univariate version $\Phi_n(\gamma) \circ x$ is well-known in the literature under the same name; see, e. g., [46]).

Looking at Eq. (3.13) it would be justified calling $\widehat{P}_{n,k}(0, X_1, \ldots, X_n)$ partial geometric polynomials. In general, let $f \in \mathcal{F}$ be any function and f_k denote its kth Taylor coefficient $D^k(f)(0)$, $k \geq 0$. Then, according to what is customary with the exponential (Bell) polynomials, we shall call $\Phi_n(\frac{f_k}{k!}\iota^k)$ *partial f-polynomials*. In just this manner we immediately obtain from Eq. (3.13) $B_{n,k}$ as partial exp-polynomials: $B_{n,k} = \Phi_n(\varepsilon_k)$, where $\varepsilon_k := \iota^k/k!$. The partial Bell polynomials itself thus prove to be FdB polynomials.

Another instance of $\widehat{P}_{n,k}$ is of interest, too. Taking $(1 + \iota)^k$ in place of ι^k we obtain

$$\Phi_n((1 + \iota)^k) = \sum_{j=0}^{n} j! \binom{k}{j} B_{n,j} = \widehat{P}_{n,k}(1, X_1, \ldots, X_n) =: P_{n,k}. \tag{3.14}$$

From this follows by virtue of binomial inversion (cf. [3, Chapter III]) the well-known formula

$$B_{n,k} = \frac{1}{k!} \sum_{j=0}^{k} (-1)^{k-j} \binom{k}{j} P_{n,j}. \tag{3.15}$$

See [22, p. 156] and [99], where Eq. (3.15) is ascribed to J. Bertrand (1864).

Remark 3.3 (Unifications).

(i) Observing Eq. (2.18) one has $\widehat{P}_{n,k} \circ 1 = P_{n,k} \circ 1 = \sum_{j=0}^{n} (k)_j (B_{n,j} \circ 1) = \sum_{j=0}^{n} (k)_j s_2(n, j) = k^n$.

(ii) Unification on both sides of Eq. (3.15) immediately yields the well-known formula $s_2(n, k) = \frac{1}{k!}\sum_{j=1}^{k}(-1)^{k-j}\binom{k}{j}j^n$ (see [84, Remark 4.1], and from a historical perspective [13, Theorem 1]).

(iii) For all $n \geq 1$ we have $\widehat{T}_n \circ 1 = T_n \circ 1 = n^{n-1} =$ the number of labeled, rooted trees on the vertex set $\{1, \ldots, n\}$.

(iv) Set $p(n) := \Phi_n(\gamma) \circ 1$. Then $p(n) = \sum_{k=0}^{n} k!s_2(n, k) =$ the number of *preferred arrangements*, that is, the total number of ordered partitions of $\{1, \ldots, n\}$ (see [85] for a Dobiński type series representing these numbers).

(v) Note that the numbers $[n]_k := \widehat{F}_{n,k} \circ 1 = F_{n,k} \circ 1$ do not agree with the falling powers $(n)_k$. Instead, we obtain from (3.11) and (3.12) $[n]_k = \sum_{j=0}^{k} s_1(k, j)j^n$ and $k^n = \sum_{j=0}^{k} s_2(k, j)[n]_j$. In [83, p. 253] it is shown that

$$[n]_k = \sum_{j=1}^{\min(k,n)} (-1)^{k-j}j!s_2(n, j)(c(k - 1, j - 1) - c(k - 1, j)).$$

A proof is also outlined in no. 14 b of the Notes and supplements to this chapter.

Proposition 3.5 (Convolution identities). *Let n, r, s be any integers, $n \geq 0$.*

(i) $\widehat{P}_{n,r+s} = \sum_{k=0}^{n} \binom{n}{k} \widehat{P}_{n-k,r}\widehat{P}_{k,s}$ (*also valid for P in place of $\widehat{P}$*),

(ii) $\binom{r + s}{r} B_{n,r+s} = \sum_{k=0}^{n} \binom{n}{k} B_{n-k,r}B_{k,s}$ $(r, s \geq 0)$.

Proof. (i). Evaluate $\widehat{P}_{n,r+s} = \widehat{\Phi}_n(\iota^r \cdot \iota^s)$ by means of the Leibniz rule (3.5). The same identity for $P_{n,r+s}$ is then obtained for $X_0 = 1$. — (ii) Replace X_0 by 0 in (i) and apply Eq. (3.13).   ◇

Remark 3.4. According to Birmajer, Gil and Weiner [10] the convolution formula (ii) seems to be established for the first time by Cvijović [24].

Proposition 3.6. $L_n = \sum_{j=1}^{n}(-1)^{j-1}\frac{1}{j}\binom{n}{j}P_{n,j}$ $(n \geq 1)$.

Proof. In the linear combination representing L_n (see Examples 3.1 (ii)) we replace $B_{n,k}$ by the right-hand side of Eq. (3.15). This gives

$$L_n = \sum_{k=1}^{n}\sum_{j=0}^{k}(-1)^{j-1}\frac{1}{k}\binom{k}{j}P_{n,j} = \sum_{j=1}^{n}\left((-1)^{j-1}P_{n,j}\underbrace{\sum_{k=j}^{n}\frac{1}{k}\binom{k}{j}}_{(*)}\right).$$

One easily verifies that $(*)$ is equal to $\frac{1}{j}\binom{n}{j}$. ◊

Remark 3.5. The statement of Proposition 3.6 is to be found in Comtet [22, p. 156], however flawed by missing the binomial factor.

Now we will give a more general version of the statements in the Propositions 3.2 and 3.3.

Theorem 3.7. *Let $f \in \mathcal{G}$ and $n, k \in \mathbb{Z}$ with $1 \le k \le n$. Then we have*

   (i) $\Omega_n(\overline{f}^k \,|\, \varphi) = k! A_{n,k}(\Omega_1(f\,|\,\varphi),\dots,\Omega_{n-k+1}(f\,|\,\varphi))$,

   (ii) $\Omega_n(\overline{f}^k \,|\, \varphi) = k!\binom{n-1}{k-1}\widehat{P}_{n-k,n}\circ\widehat{R}_\sharp\left(\frac{\Omega_1(f\,|\,\varphi)}{1},\frac{\Omega_2(f\,|\,\varphi)}{2},\dots\right)$.

Proof. (i). Observing $\overline{f}(0) = 0$ and applying Eq. (3.6) to $\overline{f}^k = \iota^k \circ \overline{f}$ we see that the left hand-side of (i) becomes $\sum_{j=0}^{n}\Omega_j(\iota^k\,|\,\varphi)\cdot(B_{n,j}\circ\Omega_\sharp(\overline{f}\,|\,\varphi))$. Clearly $\Omega_j(\iota^k\,|\,\varphi) = D^j(\iota^k)(0) = \delta_{jk}k!$. We use Proposition 3.2 to evaluate $\Omega_\sharp(\overline{f}\,|\,\varphi)$ and get the asserted by Eq. (2.14).

(ii). We proceed in a way similar to that in the argument leading to Proposition 3.3. Let g denote the function $\iota/f \in \mathcal{F}_1$. We start from the general Lagrange inversion formula $n[x^n]\overline{f}(x)^k = k[x^{n-k}]g(x)^n$ [95, Theorem 5.4.2], which may also be written as $\frac{1}{n!}D^n(\overline{f}^k)(0) = \frac{k}{n(n-k)!}D^{n-k}(g^n)(0)$. Transforming now the Taylor coefficients herein into Ω_n-terms according to Definition 3.1, we obtain by Eq. (3.7) (1-case) and by Eq. (3.10)

$$\Omega_n(\overline{f}^k\,|\,\varphi) = \frac{k(n-1)!}{(n-k)!}\Omega_{n-k}(\iota^n\circ g\,|\,\varphi)$$

$$= k!\binom{n-1}{k-1}\sum_{j=0}^{n-k}D^j(\iota^n)(\Omega_0(g\,|\,\varphi))\cdot(B_{n,j}\circ\Omega_\sharp(g\,|\,\varphi))$$

$$= k!\binom{n-1}{k-1}\widehat{P}_{n-k,n}(\Omega_0(g\,|\,\varphi),\dots,\Omega_{n-k}(g\,|\,\varphi)).$$

By Proposition 3.1 we have $\Omega_j(g \,|\, \varphi) = \widehat{R}_j(\Omega_0(\frac{f}{\iota} \,|\, \varphi), \ldots, \Omega_j(\frac{f}{\iota} \,|\, \varphi))$. Now observe (as in the proof of Corollary 3.4) that f/ι has Taylor coefficients $D^{r+1}(f)(0)/(r+1)$, whence $\Omega_r(\frac{f}{\iota} \,|\, \varphi) = \frac{1}{r+1}\Omega_{r+1}(f \,|\, \varphi)$ for $r \geq 0$. This completes the proof. $\Diamond$

Corollary 3.8. *For all* $n \geq k \geq 1$

$$A_{n,k} = \binom{n-1}{k-1} \widehat{P}_{n-k,n}(\widehat{R}_0(\tfrac{X_1}{1}), \widehat{R}_1(\tfrac{X_1}{1}, \tfrac{X_2}{2}), \ldots, \widehat{R}_{n-1}(\tfrac{X_1}{1}, \ldots, \tfrac{X_n}{n})).$$

Proof. Take $f = \varphi$ and recall that $\Omega_j(\varphi \,|\, \varphi) = X_j$. $\Diamond$

Remark 3.6. Unification on both sides of the remarkable identity above enables us to represent the Stirling numbers of the first kind by means of the potential polynomials. Define $\rho_s := \widehat{R}_s(1, \frac{1}{2}, \ldots, \frac{1}{s+1}) \in \mathbb{Q}$, $s \geq 0$; then $\rho_0 = 1$ and by Eq. (2.17)

$$s_1(n,k) = \binom{n-1}{k-1} P_{n-k,n}(\rho_1, \ldots, \rho_{n-1}) \qquad (1 \leq k \leq n).$$

We note here (without proof) that $\rho_s = \sum_{j=0}^{s}(-1)^j \binom{s+j}{s}^{-1} \widetilde{s}_2(s+j, j)$ and $\rho_{2j+1} = 0$ $(j \geq 1)$, where $\widetilde{s}_2(s+j, j) := \widetilde{B}_{s+j,j} \circ 1$ (associated Stirling numbers of the second kind; see, e. g., [77, p. 76], [22, p. 222]).

We complete the picture by also establishing statements for $\Omega_n(f^k \,|\, \varphi)$ and $B_{n,k}$ that correspond to those of Theorem 3.7 and Corollary 3.8, respectively.

Theorem 3.9. *Let* $f \in \mathcal{G}$ *and* $n, k \in \mathbb{Z}$ *with* $1 \leq k \leq n$. *Then we have*

(i) $\Omega_n(f^k \,|\, \varphi) = k! B_{n,k}(\Omega_1(f \,|\, \varphi), \ldots, \Omega_{n-k+1}(f \,|\, \varphi))$,

(ii) $\Omega_n(f^k \,|\, \varphi) = k! \binom{n}{k} \widehat{P}_{n-k,k}(\tfrac{\Omega_1(f \,|\, \varphi)}{1}, \tfrac{\Omega_2(f \,|\, \varphi)}{2}, \ldots)$.

Proof. The proof runs in much the same way as it does for Theorem 3.7. (i) Apply Eq. (3.6) to $\Omega_n(\iota^k \circ f \,|\, \varphi)$. — (ii) Choose $g \in \mathcal{F}_1$ such that $f = \iota \cdot g$; hence $f^k = \iota^k \cdot g^k$ and obviously $\Omega_n(f \,|\, \varphi) = n\Omega_{n-1}(g \,|\, \varphi)$. Then, the Leibniz rule (3.5) and a straightforward calculation give $\Omega_n(f^k \,|\, \varphi) = k! \binom{n}{k} \widehat{P}_{n-k,k} \circ \Omega_\sharp(g \,|\, \varphi)$. From this the assertion follows. $\Diamond$

Corollary 3.10. *For all* $n \geq k \geq 1$

$$B_{n,k} = \binom{n}{k} \widehat{P}_{n-k,k}\left(\tfrac{X_1}{1}, \ldots, \tfrac{X_{n-k+1}}{n-k+1}\right).$$

Proof. Take $f = \varphi$ and recall that $\Omega_j(\varphi \,|\, \varphi) = X_j$. $\Diamond$

4 Composition rules

In this section we are going to investigate the effect the composition of functions has on polynomials, which depend in a specific way on those functions. The two main results (Theorem 4.6 and Theorem 4.7) will prove to be efficient tools for dealing with the polynomial families of interest here.

Let f, g be any functions such that $h = f \circ g$ is well-defined as a function. We write $F_n = \Phi_n(f)$, $G_n = \Phi_n(g)$, $H_n = \Phi_n(h)$ for the corresponding FdB polynomials according to Eq. (3.9).

Proposition 4.1. $\Phi_n(f \circ g) = \Phi_n(f) \circ \Phi_\sharp(g).$

Proof. Recall that $\varphi \in \mathcal{F}_0$ does not occur in f or in g. In the 0-case ($f \in \mathcal{F}$, $g \in \mathcal{F}_0$) we obtain by Eq. (3.6)

$$\Phi_n(f \circ g) = \Omega_n(f \circ (g \circ \varphi) \,|\, \varphi)$$

$$= \sum_{k=0}^{n} \Omega_k(f \,|\, \varphi) \cdot (B_{n,k} \circ \Omega_\sharp(g \circ \varphi \,|\, \varphi)) \qquad (*)$$

$$= \left(\sum_{k=0}^{n} D^k(f)(0) B_{n,k} \right) \circ \Phi_\sharp(g) = \Phi_n(f) \circ \Phi_\sharp(g).$$

In the 1-case ($f \in \widehat{\mathcal{P}}$, $g \in \mathcal{F}_1$) we have to apply Eq. (3.7) so that $\Omega_k(f \,|\, \varphi)$ in line (*) becomes $D^k(f)(\Omega_0(g \circ \varphi \,|\, \varphi)) = D^k(f)(\Phi_0(g))$, and hence

$$\Phi_n(f \circ g) = \left(\sum_{k=0}^{n} D^k(f)(X_0) B_{n,k} \right) \circ \Phi_\sharp(g) = \Phi_n(f) \circ \Phi_\sharp(g). \qquad \Diamond$$

Remark 4.1. The statement of Proposition 4.1 may alternatively be written as $H_n = F_n(G_1, \ldots, G_n)$ (0-case) or as $H_n = F_n(G_0, G_1, \ldots, G_n)$ (1-case). Note that the 0-case includes $H_0 = F_0 = f(0)$, whereas in the 1-case we have $H_0 = F_0(G_0)$ with $G_0 = g(0) \neq 0$.

Remark 4.2. Suppose $f \in \mathcal{P}$ and $g \in \widehat{\mathcal{P}}$. Then we have for every $n \geq 0$
$\widehat{\Phi}_n(f \circ g) = \widehat{\Phi}_n(f) \circ \widehat{\Phi}_\sharp(g)$ (by applying the same argument as in the 1-case
in the proof of Proposition 4.1).

From Remark 4.2 we obtain useful *multiplication rules* for the potential
polynomials and the partial Bell polynomials.

Corollary 4.2 (Remark 4.2). *For all* $n \geq 0$ *and* $r, s \in \mathbb{Z}$ *we have*

$$\text{(i)} \quad \widehat{P}_{n,rs} = \widehat{P}_{n,r} \circ \widehat{P}_{\sharp,s} = \widehat{P}_{n,r}(\widehat{P}_{0,s}, \ldots, \widehat{P}_{n,s}),$$

$$\text{(ii)} \quad \widehat{P}_{n,-r} = \widehat{P}_{n,r} \circ \widehat{R}_\sharp = \widehat{P}_{n,r}(\widehat{R}_0, \ldots, \widehat{R}_n),$$

$$\text{(iii)} \quad \widehat{R}_n \circ \widehat{R}_\sharp = \widehat{R}_n(\widehat{R}_0, \ldots, \widehat{R}_n) = X_n,$$

$$\text{(iv)} \quad B_{n,rs} = \frac{r!(s!)^r}{(rs)!} B_{n,r}(B_{1,s}, \ldots, B_{n-r+1,s}) \quad (r, s \geq 0).$$

Proof. (i) $\widehat{P}_{n,rs} = \widehat{\Phi}_n(\iota^{r \cdot s}) = \widehat{\Phi}_n(\iota^r \circ \iota^s) = \widehat{\Phi}_n(\iota^r) \circ \widehat{\Phi}_\sharp(\iota^s) = \widehat{P}_{n,r} \circ \widehat{P}_{\sharp,s}.$ —
(ii) In (i) set $s = -1$. — (iii) Take $r = -1$ in (ii) and observe $\widehat{P}_{n,-1} = \widehat{R}_n$,
$\widehat{P}_{n,1} = X_n$. — (iv) Put $X_0 = 0$ in (i); then apply Eq. (3.13) and the homogeneity of the $B_{n,r}$.   ◊

Example 4.1. Let $f \in \mathcal{G}$ and $g = \overline{f}$. Then Proposition 4.1 immediately provides a quite simple infinite scheme of inverse relations. We have $F_n \circ G_\sharp =
G_n \circ F_\sharp = \Phi_n(\iota) = X_n$. For instance, $B_n \circ L_\sharp = L_n \circ B_\sharp = X_n$, where
$B_n = \Phi_n(\varepsilon)$ and $L_n = \Phi_n(\lambda)$ (see Examples 3.1). Chou, Hsu and Shiue [19]
discuss this and some more examples, which fit into this scheme.

Remark 4.3. For $n \geq 1$, let $\Phi_n[\mathcal{G}]$ denote the set of $\Phi_n(g)$, $g \in \mathcal{G}$. This set
forms (together with $\circ$) a non-abelian group. According to Proposition 4.1
we have: $\Phi_n[\mathcal{G}]$ is closed under $\circ$, and its identity is X_n, since $G_n \circ X_\sharp =
X_n \circ G_\sharp = G_n$. The inverse of $G_n = \Phi_n(g)$ is $\overline{G}_n := \Phi_n(\overline{g})$, which by
Remark 2.4, (3.9) and (2.11) can be written somewhat more explicitly as $\overline{G}_n =
\sum_{k=1}^{n} A_{k,1}^g(0) B_{n,k}$.

It should be noticed here that the mapping, which assigns to each $g \in \mathcal{G}$
the infinite sequence $(\Phi_1(g), \Phi_2(g), \Phi_3(g), \ldots)$, is an isomorphism from the
group $(\mathcal{G}, \circ)$ to the direct product $\prod_{n \geq 1} \Phi_n[\mathcal{G}]$ (in essence, the group of *formal diffeomorphisms* leaving 0 fixed; it figures as a non-commutative Hopf
algebra in [16]).

Lemma 4.3 (Substitution Lemma). *If $f \in \mathcal{F}$ and $g \in \mathcal{F}_0$, then*

(i) $\quad B_{n,k} \circ G_\sharp = B_{n,k}(G_1, \ldots, G_{n-k+1}) = \displaystyle\sum_{j=k}^{n} B_{j,k}^{g}(0) B_{n,j},$

(ii) $\quad F_n \circ G_\sharp = F_n(G_1, \ldots, G_n) = \displaystyle\sum_{j=0}^{n} F_j^{g}(0) B_{n,j}.$

Proof. (i). By Proposition 4.1 we have

$$B_{n,k} \circ G_\sharp = \Phi_n(\varepsilon_k) \circ \Phi_\sharp(g) = \Phi_n(\varepsilon_k \circ g)$$
$$= \sum_{j=0}^{n} D^j(\varepsilon_k \circ g)(0) B_{n,j}.$$

The FdB formula (2.12) yields

$$D^j(\varepsilon_k \circ g)(0) = \sum_{i=0}^{j} D^i(\varepsilon_k)(g(0)) B_{j,i}^{g}(0)$$
$$= \sum_{i=0}^{j} \delta_{ik} B_{j,i}^{g}(0) = B_{j,k}^{g}(0).$$

Since $B_{j,k}^{g}(0) = 0$ for $j < k$, this proves (i).

(ii). By Eq. (3.9) $F_n = \sum_{k=0}^{n} D^k(f)(0) B_{n,k}$, hence

$$F_n(G_1, \ldots, G_n) = \sum_{k=0}^{n} D^k(f)(0) B_{n,k}(G_1, \ldots, G_{n-k+1})$$
$$\underset{\text{(i)}}{=} \sum_{k=0}^{n} D^k(f)(0) \sum_{j=k}^{n} B_{j,k}^{g}(0) B_{n,j},$$
$$= \sum_{j=0}^{n} \left(\sum_{k=0}^{j} D^k(f)(0) B_{j,k}^{g}(0) \right) B_{n,j}.$$

The inner sum is equal to $F_j^{g}(0)$. $\qquad \lozenge$

Corollary 4.4.

(i) $\quad B_{n,k}(B_1, \ldots, B_{n-k+1}) = \displaystyle\sum_{j=k}^{n} s_2(j,k) B_{n,j},$

$$\text{(ii)} \quad B_{n,k}(L_1, \ldots, L_{n-k+1}) = \sum_{j=k}^{n} s_1(j, k) B_{n,j}.$$

Proof. (i). Put $g = \varepsilon$ in part (i) of Lemma 4.3; it follows $B_{j,k}^{\varepsilon}(0) = B_{j,k} \circ 1 = s_2(j, k)$. $-$ (ii). Substitute λ for g; hence by Eq. (2.11) $B_{j,k}^{\lambda}(0) = A_{j,k}^{\varepsilon}(0) = A_{j,k} \circ 1 = s_1(j, k)$. $\diamond$

Remark 4.4. Let $b(n)$ denote the nth *Bell number* (total number of partitions of an n-set). We then have $b(n) = B_n \circ 1$ and by part (i) of Corollary 4.4 the following identity established by Yang [105, Equation (31)]:

$$B_{n,k}(b(1), \ldots, b(n - k + 1)) = \sum_{j=k}^{n} s_2(n, j) s_2(j, k).$$

The right-hand side of (i) in Corollary 4.4 can assume yet another form by substituting for $s_2(j, k)$ the explicit expression from part (ii) in Remark 3.3. By the homogeneity of the Bell polynomials one easily obtains after a short calculation

$$B_{n,k}(B_1, \ldots, B_{n-k+1}) = \frac{1}{k!} \sum_{j=1}^{k} (-1)^{k-j} \binom{k}{j} B_n(jX_1, \ldots, jX_n). \quad (4.1)$$

Proposition 4.5. $P_{n,k}(B_1, \ldots, B_n) = B_n(kX_1, \ldots, kX_n)$.

Proof. Again using part (i) of Corollary 4.4 yields

$$P_{n,k}(B_1, \ldots, B_n) = \sum_{j=0}^{n} (k)_j B_{n,j}(B_1, \ldots, B_n) = \sum_{j=0}^{n} (k)_j \sum_{r=j}^{n} s_2(r, j) B_{n,r}$$

$$= \sum_{r=0}^{n} \left(\sum_{j=0}^{r} (k)_j s_2(r, j) \right) B_{n,r} = \sum_{r=0}^{n} k^r B_{n,r}$$

$$= \sum_{r=0}^{n} B_{n,r}(kX_1, kX_2, \ldots) = B_n(kX_1, \ldots, kX_n). \qquad \diamond$$

Let us return to Proposition 4.1, since we are now in a position to prove that also the converse statement holds. We give it a slightly different form.

Theorem 4.6 (First Composition Rule). *Let f and g be any functions such that $f \circ g \in \mathcal{F}$. Then, for all $h \in \mathcal{F}$:* $h = f \circ g \iff H_n = F_n \circ G_\sharp$.

Proof. '⇒': By Proposition 4.1. — '⇐': First assume the 0-case: $f, h \in \mathcal{F}$ and $g \in \mathcal{F}_0$. Let n be any non-negative integer and suppose $H_n = F_n \circ G_\sharp$. Part (ii) of Lemma 4.3 then yields

$$\sum_{k=0}^{n} D^k(h)(0) B_{n,k} = \sum_{k=0}^{n} F_k^g(0) B_{n,k}.$$

If $n = 0$, then $h(0) = f(0) = (f \circ g)(0)$ according to Remark 4.1. For $n > 0$ we obtain from Proposition 2.2 that the sequence $B_{n,1}, B_{n,2}, \dots, B_{n,n}$ is linearly independent in $\mathcal{K}[X_1, \dots, X_n]$. Hence for every k, $1 \le k \le n$, by the FdB formula (2.12)

$$D^k(h)(0) = F_k^g(0) = \sum_{j=0}^{k} D^j(f)(0) B_{k,j}^g(0) = D^k(f \circ g)(0). \qquad (4.2)$$

This shows that the Taylor coefficients of h agree with those of $f \circ g$. — The same reasoning works for the 1-case ($f \in \widehat{\mathcal{P}}$, $g \in \mathcal{F}_1$, $h \in \mathcal{F}$), however with $D^k(h)(X_0) = D^k(f \circ g)(X_0)$ for $k = 0, 1, 2, \dots$ instead of (4.2) at the end. Here already $k = 0$ yields the desired result. ◇

Our second main result concerns the mapping $P \mapsto P^\varphi$ in its effect on the Stirling polynomials.

Theorem 4.7 (Second Composition Rule).

$$\text{(i)} \quad B_{n,k}^{f \circ g}(0) = \sum_{j=k}^{n} B_{n,j}^g(0) B_{j,k}^f(0) \qquad (f, g \in \mathcal{F}_0),$$

$$\text{(ii)} \quad A_{n,k}^{f \circ g}(0) = \sum_{j=k}^{n} A_{n,j}^f(0) A_{j,k}^g(0) \qquad (f, g \in \mathcal{G}).$$

Remark 4.5. Part (i) of the theorem is essentially due to Jabotinsky [39, 41]. Note that the right-hand side of (i) can be interpreted as the contravariant product of two instances of the matrix $(B_{n,k})$. Comtet used this idea in order to generalize Faà di Bruno's formula so as to apply to fractionary iterates of formal series [22, p. 144].

Proof. (i). The result can be obtained by direct computation:

$$B_{n,k}^{f \circ g}(0) = B_{n,k}(D^1(f \circ g)(0), \dots, D^{n-k+1}(f \circ g)(0))$$

$$= B_{n,k}(\Phi_1(f)^g(0), \ldots, \Phi_{n-k+1}(f)^g(0)) \quad \text{(Eq. (2.12) and (3.9))}$$
$$= B_{n,k}(\Phi_1(f), \ldots, \Phi_{n-k+1}(f))^g(0). \qquad\qquad \text{(Eq. (2.9))}$$

Applying Lemma 4.3 (i) to $B_{n,k}(\Phi_1(f), \ldots, \Phi_{n-k+1}(f))$ yields the assertion.
(ii). We use part (i) and Eq. (2.11). Note the covariant behavior of $A_{n,k}$.

$$A_{n,k}^{f \circ g}(0) = (B_{n,k}^{\overline{f \circ g}} \circ (f \circ g))(0) = B_{n,k}^{\overline{g} \circ \overline{f}}(f(g(0)))$$

$$= B_{n,k}^{\overline{g} \circ \overline{f}}(0) = \sum_{j=k}^{n} B_{n,j}^{\overline{f}}(0) B_{j,k}^{\overline{g}}(0)$$

$$= \sum_{j=k}^{n} A_{n,j}^{f}(\overline{f}(0)) A_{j,k}^{g}(\overline{g}(0)).$$

Since $\overline{f}(0) = \overline{g}(0) = 0$, the result follows. $\qquad\qquad \diamond$

Corollary 4.8. *If $H_n = \Phi_n(h)$, $h \in \mathcal{F}$, then*

$$H_n^{f \circ g}(0) = \sum_{k=0}^{n} B_{n,k}^{g}(0) H_k^{f}(0).$$

Proof. By a short calculation using Eq. (3.9) and Theorem 4.7 (i). $\qquad \diamond$

The following statement was originally published in [84, Theorem 5.1] under the title 'Inversion Law' and was proven there by induction. The new proof presented below is much more natural.

Corollary 4.9. *$(A_{n,k})$ and $(B_{n,k})$ are orthogonal companions of each other:*
$A_{n,k} = B_{n,k}^{\perp}$ and $B_{n,k} = A_{n,k}^{\perp}$.

Proof. Let φ be any function from $\mathcal{G}$. Then

$$B_{n,k}^{\varphi \circ \overline{\varphi}}(0) = B_{n,k}^{\iota}(0) = B_{n,k}(1, 0, \ldots, 0) = \delta_{nk}. \qquad (*)$$

On the other hand, by part (i) of Theorem 4.7 and by Eq. (2.9) we have

$$B_{n,k}^{\varphi \circ \overline{\varphi}}(0) = \sum_{j=k}^{n} A_{n,j}^{\varphi}(\overline{\varphi}(0)) B_{j,k}^{\varphi}(0) = \left(\sum_{j=k}^{n} A_{n,j} B_{j,k} \right)^{\varphi}(0). \qquad (**)$$

Now equate the right-hand sides of (*) and (**). Then, finally applying the argument from Remark 2.3 shows that the orthogonality relation Eq. (2.10) is satisfied by $A_{n,k}$ and $B_{n,k}$. $\qquad\qquad \diamond$

5 Representation by Bell polynomials

As we have seen in Section 4, composing FdB polynomials again yields FdB polynomials. In this section we will go beyond by dealing with polynomial families (not necessarily FdB) that can be represented as instances of the partial Bell polynomials. A well-known example is the cycle indicator. In addition, several new polynomial families will be introduced and examined in more detail, including 'forest polynomials' (generated by tree polynomials) as well as multivariate Lah polynomials, which form a self-orthogonal family. Since a (regular) family represented this way always has an orthogonal companion, we can establish a corresponding inverse relation in each case.

5.1 B-representability

A polynomial family $(Q_{n,k})$ is said to be *B-representable*, if there is an infinite sequence of polynomials $H_1, H_2, H_3, \ldots$ such that for all integers n, k with $1 \le k \le n$

$$Q_{n,k} = B_{n,k} \circ H_\sharp = B_{n,k}(H_1, \ldots, H_{n-k+1}). \tag{5.1}$$

The Bell polynomials itself are, of course, B-representable (since $B_{n,k} = B_{n,k} \circ B_{\sharp,1}$ with $B_{n,1} = X_n$). The same holds for the associate Bell polynomials $\widetilde{B}_{n,k}$ [84, Corollary 4.5] with $\widetilde{B}_{j,1} = X_j$ for $j \ge 2$, but $\widetilde{B}_{1,1} = 0$ (implying that this family is not regular). Non-trivial examples of B-representable families provide $(A_{n,k})$ (see Eq. (2.14)), or the Stirling numbers $s_2(n,k)$ (see Eq. (2.18)).

First, we gather some basic properties.

Proposition 5.1. *Let $(Q_{n,k})$ be any B-representable family of polynomials. Then we have*

(i) *$(Q_{n,k})$ is lower triangular.*

(ii) *$Q_{n,k} = B_{n,k}(Q_{1,1}, \ldots, Q_{n-k+1,1})$ $(1 \le k \le n)$*

(iii) *$Q_{n,n} = (Q_{1,1})^n$*

(iv) *If $(Q_{n,k})$ is regular, then $Q_{n,k}^\perp = A_{n,k}(Q_{1,1}, \ldots, Q_{n-k+1,1})$ exists and is B-representable.*

Proof. (i) Clear by definition. — (ii) Taking $k = 1$ in Eq. (5.1) gives $Q_{n,1} = B_{n,1}(H_1, \ldots, H_n) = H_n$. — (iii) A special case of (ii) is $Q_{1,1} = H_1$; hence,

putting $k = n$ in Eq. (5.1) we get $Q_{n,n} = B_{n,n}(H_1) = (H_1)^n = (Q_{1,1})^n$. —
(iv) It follows from regularity and triangularity $Q_{j,j} \neq 0$ for all j (diagonal entries), in particular $Q_{1,1} \neq 0$. Therefore, $A_{n,k}(Q_{1,1}, \ldots, Q_{n-k+1,1})$ is well-defined (see Eq. (2.16)) and by Corollary 4.9 equal to $Q_{n,k}^\perp$. From this we finally obtain by Eq. (2.14)

$$B_{n,k}(Q_{1,1}^\perp, Q_{2,1}^\perp, \ldots, Q_{n-k+1,1}^\perp)$$
$$= B_{n,k}(A_{1,1}(Q_{1,1}), A_{2,1}(Q_{1,1}, Q_{2,1}, \ldots), \ldots)$$
$$= A_{n,k}(Q_{1,1}, Q_{2,1}, \ldots, Q_{n-k+1,1}) = Q_{n,k}^\perp. \qquad \Diamond$$

Part (iii) of Proposition 5.1 may serve as a *necessary condition for B-representability*. Since, for example, $P_{1,1} = X_1$ and $P_{2,2} = 2X_1^2 + 2X_2 \neq (P_{1,1})^2$, the potential polynomials are not B-representable.

Remark 5.1. As a precaution, it should be emphasized that criterion (iii) is not sufficient. The crucial point here is that Eq. (5.1) comprises an *infinite* number of equations and unknowns $H_1, H_2, H_3, \ldots$. On the other hand: Given any fixed $n \geq 1$, the system consisting of the first n equations from (5.1) is indeed solvable, if and only if there exists H_1 such that $(H_1)^n = Q_{n,n}$.

To see this, recall that $H_n = Q_{n,1}$ and observe that the remaining $H_2, \ldots, H_{n-1}$ appear only linearly, since Eq. (2.20) yields

$$\frac{\partial B_{n,k}}{\partial X_{n-k+1}}(H_1, \ldots, H_{n-k+1}) = \binom{n}{k-1} H_1^{k-1}$$

for every k with $2 \leq k \leq n - 1$.

As an example, put $n = 3$ and consider the equation system

$$B_{3,k}(H_1, \ldots, H_{4-k}) = Q_{3,k} := 1, \quad 1 \leq k \leq 3.$$

If we choose $H_1 = 1$, then necessarily $H_2 = 1/3$ and $H_3 = 1$. But this solution cannot be extended to $n = 4$, for choosing in this case $H_1 = 1$ implies $H_2 = 1/6$, $H_3 = -11/48$, and $H_4 = 1$ so that there is no infinite sequence $H_1, H_2, H_3, \ldots$ satisfying $B_{n,k}(H_1, \ldots, H_{n-k+1}) = 1$ *for all* n, k with $1 \leq k \leq n$.

We will now characterize the B-representable polynomial families. As to FdB polynomials, the following can be easily inferred from the First Composition Rule (FCR).

Proposition 5.2. *Let $(Q_{n,k})$ be any family of FdB polynomials. Then we have: $(Q_{n,k})$ is B-representable $\iff Q_{n,k} = \Phi_n(\frac{h^k}{k!})$ for some $h \in \mathcal{F}$.*

Proof. '⇒': Let f_k be functions such that $\Phi_n(f_k) = Q_{n,k} = B_{n,k} \circ Q_{\sharp,1}$. According to the FCR (Theorem 4.6, '⇐') the equation $f_k = \varepsilon_k \circ f_1 = f_1^k/k!$ holds, where $f_1 \in \mathcal{F}.$ — '⇐': Immediately, by the FCR ('⇒') $Q_{n,k} = \Phi_n(\frac{h^k}{k!}) = \Phi_n(\varepsilon_k \circ h) = \Phi_n(\varepsilon_k) \circ \Phi_\sharp(h) = B_{n,k} \circ Q_{\sharp,1}$. $\Diamond$

In the case of an arbitrary polynomial family $(Q_{n,k})$, the statement to follow provides a necessary and sufficient condition for B-representability.

Proposition 5.3. $(Q_{n,k})$ *is B-representable, if and only if*

$$Q_{n,k} = \sum_{j=1}^{n-k+1} \binom{n-1}{j-1} Q_{j,1} Q_{n-j,k-1}. \tag{*}$$

Proof. Necessity (›only if‹): Immediate by the fact that $Q_{n,k} := B_{n,k}$ satisfies (*); see, e. g., [49] and [84, Proposition 5.5, Remark 5.6](cf. Chapter I). Sufficiency (by induction): Suppose (*); then, observing

$$Q_{j,1} = B_{j,1}(Q_{1,1}, \ldots, Q_{j,1})$$

and applying the induction hypothesis

$$Q_{n-j,k-1} = B_{n-j,k-1}(Q_{1,1}, \ldots, Q_{n-k-j+2,1}) \quad (1 \le j \le n-k+1)$$

one gets

$$Q_{n,k} = \sum_{j=1}^{n-k+1} \binom{n-1}{j-1} B_{j,1}(Q_{1,1}, Q_{2,1} \ldots) B_{n-j,k-1}(Q_{1,1}, Q_{2,1}, \ldots)$$

$$= B_{n,k}(Q_{1,1}, \ldots, Q_{n-k+1,1}) \quad \text{(since } B_{n,k} \text{ satisfies (*))}. \qquad \Diamond$$

5.2 Generalized Stirling inversion

The ordinary Stirling inversion [94, Proposition 1.4.1b] is based on the well-known orthogonality relation satisfied by the Stirling numbers of the first and second kind: $\sum_{j=k}^{n} s_1(n,j) s_2(j,k) = \delta_{nk}$ for all $n \ge 0$ and $0 \le k \le n$. This type of inversion can be generalized considerably by taking advantage of the fact that according to Proposition 5.1 (iv), every regular B-representable family of polynomials $Q_{n,k}$ has an orthogonal companion $Q_{n,k}^{\perp}$, which is also B-representable.

Proposition 5.4 (Generalized Stirling inversion). *Let $U_0, U_1, U_2, \ldots$ and $V_0, V_1, V_2, \ldots$ be two sequences (of polynomials from an arbitrary overring of $\mathcal{K}[X_1^{-1}, X_1, X_2, \ldots]$, say) and let $(Q_{n,k})$ be any regular B-representable family of polynomials. Then the following statements are equivalent:*

$$\text{(i)} \quad U_n = \sum_{k=0}^{n} Q_{n,k} V_k \quad \text{for all } n \geq 0,$$

$$\text{(ii)} \quad V_n = \sum_{k=0}^{n} Q_{n,k}^{\perp} U_k \quad \text{for all } n \geq 0.$$

Proof. By Corollary 4.9 $B_{n,k}^{\perp} = A_{n,k}$; hence (i) and (ii) are equivalent when $Q_{n,k}$ is replaced by $B_{n,k}$ (and consequently $Q_{n,k}^{\perp}$ by $A_{n,k}$); see [84, Corollary 5.2]. On the other hand, we have by assumption $Q_{n,k} = B_{n,k} \circ Q_{\sharp,1}$, and thus by part (iv) of Proposition 5.1 also $Q_{n,k}^{\perp} = A_{n,k} \circ Q_{\sharp,1}$. From this follows the assertion. $\diamond$

Remark 5.2. The above generalizations by no means exhaust all possibilities of basing inverse relationships on orthogonality. In principle, any invertible infinite matrix $(Q_{n,k})$ can be used for this purpose (for convenience, of lower triangular form). For example, Riordan [78, p. 100] compiled a large stock of inverse relations by establishing identities involving the Taylor coefficients of a function $f \in \mathcal{F}_1$ and its reciprocal f^{-1}. With the concepts developed in Section 3, it is easy to convince oneself that all these relations fall under the orthogonality scheme

$$\sum_{j=k}^{n} \frac{X_{n-j}}{(n-j)!} \frac{\widehat{R}_{j-k}}{(j-k)!} = \delta_{nk}.$$

For the general case, Milne and Bhatnagar [70] have found recurrences, which characterize the entries $Q_{n,k}$ of an orthogonality relation. Huang [38], by representing functions relative to strictly monotone Schauder bases in $\mathcal{F}$, has shown that the inverse relationship (with the orthogonality property) comes about through the interchange of bases.

While the matrix entries $Q_{n,k}$ in the orthogonality relation generally only determine each other implicitly, in the case of B-representable polynomials we immediately have the explicit expression $Q_{n,k}^{\perp} = A_{n,k} \circ Q_{\sharp,1}$. We can go a step further here and resolve $A_{n,k}$ into lower-order expressions, for example according to Corollary 3.8, to Theorem 2.3, or to the general polynomial version of the famous Schlömilch formula for the Stirling numbers $s_1(n, k)$ in terms of $s_2(n, k)$. This generalization has been presumably for the first time established and proven in [84, Theorem 6.4]. After some few index shifts and applying elementary properties of the binomial coefficients, the latter result can be easily rewritten in the form of an identity of the following Schlömilch-Schläfli type:

$$A_{n,n-k} = (-1)^k \sum_{j=0}^{k} \binom{k+n}{k-j} \binom{k-n}{k+j} X_1^{-(n+j)} B_{k+j,j}. \qquad (5.2)$$

By finally applying Proposition 5.1 to this, we are easily led to the corresponding generalizations regarding B-representable polynomials.

Theorem 5.5 (Generalized Schlömilch-Schläfli identities). *For every regular B-representable family of polynomials $Q_{n,k}$ the following holds for all $n \geq k \geq 0$:*

(i) $\quad Q_{n,n-k}^{\perp} = (-1)^k \sum_{j=0}^{k} \binom{k+n}{k-j} \binom{k-n}{k+j} (Q_{1,1}^{\perp})^{n+j} Q_{k+j,j},$

(ii) $\quad Q_{n,n-k} = (-1)^k \sum_{j=0}^{k} \binom{k+n}{k-j} \binom{k-n}{k+j} (Q_{1,1})^{n+j} Q_{k+j,j}^{\perp}.$

Remark 5.3. Note that $Q_{1,1}^{\perp} = A_{1,1}(Q_{1,1}) = Q_{1,1}^{-1}$.

Remark 5.4. Unification on both sides of (5.2) immediately yields Schläfli's formula for $s_1(n, n - k)$ in terms of Stirling numbers of the second kind (see Quaintance and Gould [76, Eq. (13.31)]). Conversely, if we take $B_{n,n-k}$ for $Q_{n,n-k}$ and apply unification to part (ii) of Theorem 5.5, we get Gould's formula for $s_2(n, n - k)$ in terms of Stirling numbers of the first kind (see [32] and [76, Eq. (13.42)]).

5.3 Cycle indicator polynomials

Let $\mathbb{P}$ be the union of all sets of partition types. We define the mapping $\zeta : \mathbb{P} \longrightarrow \mathbb{N}$ by

$$\zeta(r_1, r_2, r_3, \ldots) := \frac{(r_1 + 2r_2 + 3r_3 + \cdots)!}{r_1! r_2! r_3! \cdots 1^{r_1} 2^{r_2} 3^{r_3} \cdots}. \qquad (5.3)$$

The right-hand side of (5.3) is Cauchy's famous expression that counts the permutations having exactly r_j cycles of size j $(j = 1, 2, 3, \ldots)$. The corresponding partition polynomial

$$Z_{n,k} := \sum_{\mathbb{P}(n,k)} \zeta(r_1, r_2, r_3, \ldots) X_1^{r_1} X_2^{r_2} X_3^{r_3} \cdots$$

could rightly be called *partial cycle indicator*, inasmuch the term *cycle indicator* [77, p. 68] (sometimes also *augmented cycle index* [95, p. 19]) is reserved for $Z_n := Z_{n,1} + Z_{n,2} + \cdots + Z_{n,n}$.

It is easy to check that $(Z_{n,k})$ is B-representable. We have $Z_{n,n} = X_1^n = Z_{1,1}^n$ (the necessary condition (iii) from Proposition 5.1) and $Z_{n,1} = (n-1)! X_n$. Thus, a few lines of a direct calculation (cf. [22, p. 247]) result in

$$B_{n,k} \circ Z_{\sharp,1} = B_{n,k}(0! X_1, 1! X_2, 2! X_3, \ldots) = Z_{n,k}. \tag{5.4}$$

By part (iv) of Proposition 5.1 we obtain as orthogonal companion $Z_{n,k}^{\perp} = A_{n,k}(0! X_1, 1! X_2, 2! X_3, \ldots)$. As is well known, unification of (5.4) gives the signless Stirling numbers of the first kind $Z_{n,k} \circ 1 = c(n,k) = B_{n,k}(0!, 1!, 2!, \ldots)$ [22, p. 135]. Hence $Z_{n,k}^{\perp} \circ 1 = A_{n,k}(0!, 1!, 2!, \ldots) = (-1)^{n-k} s_2(n,k)$, which might be called *signed* Stirling numbers of the second kind.

Remark 5.5. From Eq. (5.4) it is only a small step to the so-called *exponential formula* that is based on the idea of interpreting the coefficients of $e^{f(x)}$ combinatorially. Here f is assumed to be any function in $\mathcal{F}_0$ and $t_n := D^n(f)(0)/(n-1)!$ for $n \geq 1$. Then, the nth Taylor coefficient of $e^{f(x)}$ is $B_n^f(0) = B_n(0! t_1, \ldots, (n-1)! t_n) = Z_n(t_1, \ldots, t_n)$, and we immediately obtain

$$\exp\left(\sum_{n \geq 1} t_n \frac{x^n}{n}\right) = \sum_{n \geq 0} Z_n(t_1, \ldots, t_n) \frac{x^n}{n!}.$$

For a detailed treatment, various combinatorial applications and historical notes on this topic, the reader is referred to Stanley [95, Section 5.1].

5.4 Idempotency polynomials. Forest polynomials

While the potential polynomials itself are not B-representable, this is actually yet the case with certain closely related families to be investigated in the sequel.

A simple example of this kind follows directly from Corollary 3.10:

$$B_{n,k}(X_0, 2X_1, 3X_2, \ldots) = \binom{n}{k} \widehat{P}_{n-k,k}. \tag{5.5}$$

This identity was established by Comtet in a slightly modified form (with $X_0 = 1$) [22, Suppl. no. 4, p. 156/7]. Observing part (i) of Remark 3.3 we immediately obtain by unification

$$B_{n,k}(1, 2, 3, \ldots) = \binom{n}{k} k^{n-k}, \tag{5.6}$$

which equals the number of idempotent maps from an n-set into itself having exactly k cycles (see, e. g., [34]). Comtet's argument is based on considering $B_{n,k}^{\psi}(0)$ with $\psi(x) := xe^x$ [22, p. 135].

Given this combinatorial meaning, it seems justified calling the expressions on the right-hand side of Eq. (5.5) *idempotency polynomials*. The following theorem exhibits them (slightly modified) in the role of an orthogonal companion.

Theorem 5.6. *For all $n, k \in \mathbb{Z}$ with $1 \leq k \leq n$ we have*

$$(i) \qquad B_{n,k}(\widehat{T}_1, \ldots, \widehat{T}_{n-k+1}) = \binom{n-1}{k-1} \widehat{P}_{n-k,n},$$

$$(ii) \qquad A_{n,k}(\widehat{T}_1, \ldots, \widehat{T}_{n-k+1}) = \binom{n}{k} \widehat{P}_{n-k,-k}.$$

Ahead of the proof three remarks are in order.

Remark 5.6. Unification on part (i) of Theorem 5.6 immediately yields

$$B_{n,k}(1^0, 2^1, 3^2, \ldots) = \binom{n-1}{k-1} n^{n-k}.$$

This numerical identity has been formulated and given a lengthy proof by Khelifa and Cherruault [47]. Abbas and Bouroubi [1, Theorems 3 and 6] provided a significantly shorter argument together with an extension to binomial sequences.

Remark 5.7. One might say, somewhat jokingly, that the Bell polynomials return forests on receiving trees. Recall that there are $\widehat{T}_n \circ 1 = n^{n-1}$ rooted (labeled) trees on n vertices. On the other hand, $\binom{n-1}{k-1} n^{n-k}$ counts the planted forests with k components on n vertices (see, e. g., [95, Proposition 5.3.2]). Motivated by this, we shall refer to the expressions on the right-hand side of (i) as *forest polynomials*, denoted by $W_{n,k}$. Of course we have $W_{j,1} = \widehat{T}_j$.

Remark 5.8. The forest polynomials form a regular family; so, according to Proposition 5.1 (iv) its orthogonal companion $W_{n,k}^{\perp}$ is the polynomials, which appear in Theorem 5.6 (ii). It differs from the idempotency polynomials only in that the second index is negated. By unification, the orthogonal relationship is immediately passed on to the corresponding number sequences (the orthogonality of which has been observed by Wang and Wang [103]).

Theorem 5.6. The proof can be carried out exclusively using polynomial identities. The key idea is here to express the tree polynomials as

$$\widehat{T}_n = A_{n,1}(\widehat{R}_0, 2\widehat{R}_1, 3\widehat{R}_2, \ldots), \tag{*}$$

which follows from Corollary 3.4 (ii) and Corollary 4.2 (iii). We then have

$$
\begin{aligned}
B_{n,k}(\widehat{T}_1, \widehat{T}_2, \widehat{T}_3, \ldots) &= B_{n,k}(A_{1,1}(\widehat{R}_0), A_{2,1}(\widehat{R}_0, 2\widehat{R}_1), \ldots)) && \text{(by (*))} \\
&= A_{n,k}(\widehat{R}_0, 2\widehat{R}_1, 3\widehat{R}_2, \ldots) && \text{(by Eq. (2.14))} \\
&= \binom{n-1}{k-1} \widehat{P}_{n-k,n}, && \text{(by Cor. 3.8, Cor. 4.2 (iii))}
\end{aligned}
$$

which proves (i). In the same manner, (ii) can be shown by applying (*), Eq. (2.15), Eq. (5.5), and Corollary 4.2 (ii):

$$A_{n,k}(\widehat{T}_1, \widehat{T}_2, \widehat{T}_3, \ldots) = B_{n,k}(\widehat{R}_0, 2\widehat{R}_1, 3\widehat{R}_2, \ldots) = \binom{n}{k} \widehat{P}_{n-k,-k}. \quad \Diamond$$

Proposition 5.7. $A_{n,k}(X_0, 2X_1, 3X_2, \ldots) = \binom{n-1}{k-1} \widehat{P}_{n-k,-n}.$

Proof. To evaluate the left-hand-side, use Corollary 3.8, and then apply Corollary 4.2 (ii). $\Diamond$

5.5 Lah polynomials. Involution

Analogous to the considerations in Section 5.3 we define the mapping $\omega : \mathbb{P} \longrightarrow \mathbb{N}$ by

$$\omega(r_1, r_2, r_3, \ldots) := \frac{(r_1 + 2r_2 + 3r_3 + \cdots)!}{r_1! r_2! r_3! \cdots}. \tag{5.7}$$

The right-hand side of Eq. (5.7) counts the number of ways a set of $n = r_1 + 2r_2 + 3r_3 + \cdots$ objects can be partitioned into linearly ordered subsets, r_j denoting the number of subsets with j elements ($j = 1, 2, 3, \ldots$). The corresponding partition polynomials

$$L_{n,k}^{+} := \sum_{\mathbb{P}(n,k)} \omega(r_1, r_2, r_3, \ldots) X_1^{r_1} X_2^{r_2} X_3^{r_3} \cdots$$

will be called *unsigned Lah polynomials*. A simple computation shows that they form a B-representable family. We have $L_{n,1}^{+} = n! X_n$ and

$$B_{n,k} \circ L_{\natural,1}^{+} = B_{n,k}(1! X_1, 2! X_2, 3! X_3, \ldots) = L_{n,k}^{+}. \tag{5.8}$$

From this it can easily be derived that $L_{n,k}^{+} \circ 1$ are the unsigned Lah numbers $l^{+}(n,k) := \frac{n!}{k!} \binom{n-1}{k-1}$ [22, p. 135].

Let us now consider the signed Lah numbers $l(n,k) := (-1)^n l^{+}(n,k)$, which are known to be self-orthogonal in the sense that $\sum_{j=k}^{n} l(n,j) l(j,k) = \delta_{nk}$; see, e.g., [77, p. 44] and [84, Examples 5.2 (ii)]. It is therefore natural to look for self-orthogonal polynomials $L_{n,k}$ such that the signed Lah numbers can be obtained by unification: $L_{n,k} \circ 1 = l(n,k)$.

At first glance, $B_{n,k}(-1! X_1, 2! X_2, -3! X_3, \ldots) = (-1)^n L_{n,k}^{+}$ might be supposed to be a suitable candidate; but this fails because these polynomials are not orthogonal companions of their own. It is however a promising (and eventually working) idea to raise to the level of multivariate polynomials the well-known identity expressing the signed Lah numbers by the Stirling numbers of the first and second kind [77, p. 44], which is mirrored in the following

Definition 5.1 (Signed Lah polynomials). $L_{n,k} := \sum_{j=k}^{n} (-1)^j A_{n,j} B_{j,k}$.

Using this definition we immediately regain by unification (and observing (2.17), (2.18)) the numerical identity

$$L_{n,k} \circ 1 = \sum_{j=k}^{n} (-1)^j s_1(n,j) s_2(j,k) = l(n,k),$$

which just has been alluded to. We also have $L_{n,k} \in \mathcal{K}[X_1^{-1}, X_2, \ldots, X_n]$ and the homogeneity $L_{n,k}(t X_1, t X_2, \ldots) = t^{-(n-k)} L_{n,k}$. The instances of $L_{5,k}$, $1 \leq k \leq 5$, may serve as an illustration:

$$L_{5,1} = -\frac{210 X_2^4}{X_1^8} + \frac{120 X_3 X_2^2}{X_1^7} - \frac{30 X_4 X_2}{X_1^6},$$

$$L_{5,2} = -\frac{270X_2^3}{X_1^6} + \frac{40X_3X_2}{X_1^5} - \frac{10X_4}{X_1^4},$$

$$L_{5,3} = -\frac{120X_2^2}{X_1^4}, \quad L_{5,4} = -\frac{20X_2}{X_1^2}, \quad L_{5,5} = -1.$$

Indeed, the remarkable orthogonality relation $L_{n,k}^\perp = L_{n,k}$ also applies.

Proposition 5.8. $\sum_{j=k}^n L_{n,j}L_{j,k} = \delta_{nk}$ $\quad (1 \le k \le n)$.

Proof. The assertion follows from Definition 5.1 by a direct straightforward computation and applying twice Corollary 4.9. $\qquad \diamond$

The main result concerning Lah polynomials provides a characterization of all regular B-representable polynomial families, which are orthogonal companions of their own.

Theorem 5.9. *Let $(Q_{n,k})$ be any regular B-representable family of polynomials. Then, $Q_{n,k}^\perp = Q_{n,k}$ holds, if and only if there exists a family of FdB polynomials $(H_1, H_2, H_3, \ldots) \in \prod_{n \ge 1} \Phi_n[\mathcal{G}]$ such that $Q_{n,k} = L_{n,k} \circ H_\sharp$.*

Proof. Sufficiency (›if‹): Immediately from Proposition 5.8 by substituting H_j for X_j $(j = 1, 2, 3, \ldots)$.

Necessity (›only if‹): Let f be any function from $\mathcal{G}$. We then define the function $\varphi(x) := \sum_{n \ge 1} Q_{n,1}^f(0)\frac{x^n}{n!}$, which is invertible, as $Q_{n,k}$ is regular. With this we obtain

$$
\begin{aligned}
B_{n,k}^\varphi(0) &= B_{n,k}(Q_{1,1}^f(0), \ldots, Q_{n-k+1,1}^f(0)) \\
&= Q_{n,k}^f(0) && \text{(by (2.9), $Q_{n,k}$ B-representable)} \\
&= (Q_{n,k}^\perp)^f(0) && \text{(by assumption)} \\
&= A_{n,k}(Q_{1,1}^f(0), \ldots, Q_{n-k+1,1}^f(0)) && \text{(by (2.9), Prop. 5.1 (iv))} \\
&= A_{n,k}^\varphi(0) = B_{n,k}^{\overline{\varphi}}(0) && \text{(by (2.11), $\varphi(0) = 0$).}
\end{aligned}
$$

By the Identity Lemma 2.4 we have $\varphi = \overline{\varphi}$. As an involutory function, φ can be written in the form $\varphi = \widetilde{g} \circ \overline{g}$, where g is an appropriate function from $\mathcal{G}$, and $\widetilde{g}(x) = g(-x)$ (see, e. g.,[60]). It follows

$$
\begin{aligned}
Q_{n,k}^f(0) = B_{n,k}^\varphi(0) &= B_{n,k}^{\widetilde{g} \circ \overline{g}}(0) \\
&= \sum_{j=k}^n B_{n,j}^{\overline{g}}(0) B_{j,k}^{\widetilde{g}}(0) && \text{(by Jabotinsky's Theorem 4.7 (i))}
\end{aligned}
$$

$$= \sum_{j=k}^{n} (-1)^j A^g_{n,j}(0) B^g_{j,k}(0) \qquad \text{(by (2.11), } \overline{g}(0) = 0)$$

$$= L^g_{n,k}(0) \qquad \text{(by Definition 5.1)}.$$

Hence, putting $h := g \circ \overline{f} \in \mathcal{G}$ and $H_m := \Phi_m(h)$ we get

$$Q^f_{n,k}(0) = L^{h \circ f}_{n,k}(0) = L_{n,k} \circ D^\sharp(h \circ f)(0)$$

$$= L_{n,k} \circ \sum_{j=0}^{\sharp} D^j(h)(0) B^f_{\sharp,j}(0) \qquad \text{(by Eq. (2.12))}$$

$$= L_{n,k} \circ \Phi_\sharp(h)^f(0)$$

$$= L_{n,k}(H_1, H_2, H_3, \ldots)^f(0) \qquad \text{(by Eq. (2.9))}.$$

Applying the argument from Remark 2.3 completes the proof. ◇

Corollary 5.10. *The signed Lah polynomials* $(L_{n,k})$ *are B-representable.*

Proof. Let h be the function from the proof of Theorem 5.9. Then, for $j = 1, 2, 3, \ldots$ set $\overline{H}_j := \Phi_j(\overline{h})$. By the FCR we have $H_j \circ \overline{H}_\sharp = X_j$. The assertion now readily follows by replacing each X_j in $L_{n,k}(H_1, H_2, \ldots) = B_{n,k}(Q_{1,1}, Q_{2,1}, \ldots)$ with $\overline{H}_j$. ◇

The Lah polynomials allow us to characterize involutory functions.

Proposition 5.11. *A function* $f \in \mathcal{G}$ *is involutory, if and only if there exists* $g \in \mathcal{G}$ *such that* $f(x) = \sum_{n \geq 1} L^g_{n,1}(0) \frac{x^n}{n!}$.

Proof. Necessity: As in the proof of Theorem 5.9 we can write $f = \widetilde{g} \circ \overline{g}$ for some $g \in \mathcal{G}$. Again it follows that $D^n(f)(0) = L^g_{n,1}(0)$.

Sufficiency: By the assumption and by Corollary 5.10

$$B^f_{n,k}(0) = B_{n,k}(L^g_{1,1}(0), \ldots, L^g_{n-k+1,1}(0)) = L^g_{n,k}(0).$$

Now, applying the FdB formula (2.12) and Proposition 5.8 we obtain for every $n \geq 1$

$$D^n(f \circ f)(0) = \sum_{k=1}^{n} D^k(f)(0) B^f_{n,k}(0) = \sum_{k=1}^{n} L^g_{n,k}(0) L^g_{k,1}(0) = \delta_{n1},$$

that is, we have $f \circ f = \iota$. ◇

From the above reasoning it can be seen immediately that, given any involution $f = \widetilde{g} \circ \overline{g}$, $g \in \mathcal{G}$, the corresponding f-polynomial $\Phi_n(f)$ takes the form

$$J_{g,n} := \sum_{k=1}^{n} L_{k,1}^{g}(0) B_{n,k}. \tag{5.9}$$

According to the FCR, the *involution polynomials* (5.9) are self-inverse, that is, they satisfy the relation $J_{g,n} \circ J_{g,\sharp} = X_n$.

5.6 Comtet's polynomials

We now return to Comtet's attempt [21] (already mentioned in Section 2.1), to determine the class of polynomials associated with the higher-order Lie operator $(\theta D)^n$, $\theta \in \mathcal{F}$. A statement concerning expansion and recurrence, analogous to Propositions 2.1 and 2.2, serves as the starting point.

Proposition 5.12. *There exist polynomials $C_{n,k} \in \mathcal{K}[X_0, \dots, X_{n-k}]$ such that*

$$(\theta D)^n = \sum_{k=0}^{n} C_{n,k}^{\theta} \cdot D^k.$$

The family $(C_{n,k})$ is triangular, regular, and uniquely determined by the differential recurrence

$$C_{n+1,k} = X_0 \left(C_{n,k-1} + \sum_{j=0}^{n-k} X_{j+1} \frac{\partial C_{n,k}}{\partial X_j} \right), \quad C_{n,0} = \delta_{n0}.$$

Proof. By a straightforward inductive argument as applied in the proof of Proposition 2.1; cf. [84, Propositions 3.1 and 3.5].   ◇

From this the following representation can be inferred:

$$C_{n,k} = \sum_{\mathbb{P}(2n-k,n)} \gamma_{n,k}(r_0, \dots, r_{n-k}) X_0^{r_0} \cdots X_{n-k}^{r_{n-k}}, \tag{5.10}$$

the sum ranging over all non-negative integral values of r_0 to r_{n-k} such that $r_0 + \cdots + r_{n-k} = n$ and $r_0 + 2r_1 + 3r_2 + \cdots = 2n - k$. The coefficients $\gamma_{n,k}(r_0, \dots, r_{n-k})$ turn out to be positive integers. In [21, Section 5] Comtet has tabulated $(C_{n,k})_{1 \le k \le n}$ up to $n = 7$ and claimed (without proof) that

$C_{n,k} \circ 1 = c(n,k)$ and $C_{n,k}(1,1,0,\ldots,0) = s_2(n,k)$. His main result [ibid., Equation (8), p. 166)] provides the following expression in diophantine form:

$$C_{n,k} = \frac{X_0}{k!} \cdot \sum_{\rho(n,1)=n-k} k \cdot \prod_{j=1}^{n-1} \frac{k + \rho(n,j) - j}{r_j!} X_{r_j}, \qquad (5.11)$$

where $\rho(n,j)$ denotes the sum $r_1 + \cdots + r_{n-j}$ (with non-negative integers $r_1, r_2, r_3, \ldots$). This formula appears only to a modest extent suitable for computational purposes. For example, although $C_{6,2}$ consists of only 5 monomials, a total of 70 solutions of the equation $\rho(6,1) = 4$ has to be checked in order to finally obtain $C_{6,2} = 31X_0^2 X_1^4 + 146 X_0^3 X_1^2 X_2 + 34 X_0^4 X_2^2 + 57 X_0^4 X_1 X_3 + 6 X_0^5 X_4$.

In the following it will be shown that Comtet's polynomial family $(C_{n,k})$ can be smoothly integrated into our algebraic framework developed so far. In particular, it turns out that $C_{n,k}$ can be represented by the Stirling polynomials of the first and second kind.

Theorem 5.13. $C_{n,k} = A_{n,k}(\widehat{R}_0, \ldots, \widehat{R}_{n-k})$ $\qquad (0 \le k \le n)$.

Proof. Let φ be any function such that $D(\varphi) \in \mathcal{F}_1$. According to Propositions 5.12 and 2.1 we have

$$(\theta D)^n = \sum_{k=0}^{n} C_{n,k}^\theta D^k \quad \text{and} \quad (D(\varphi)^{-1} D)^n = \sum_{k=0}^{n} A_{n,k}^\varphi D^k. \qquad (*)$$

Choose $\theta = D(\varphi)^{-1}$, so that both expansions agree. Hence, by Proposition 3.1 we obtain for every $j \ge 1$

$$D^j(\varphi) = D^{j-1}(\theta^{-1}) = \widehat{R}_{j-1}^\theta$$

and by equating the coefficients of D^k in (*)

$$C_{n,k}^\theta = A_{n,k}^\varphi = A_{n,k} \circ D^\sharp(\varphi) = A_{n,k}(\widehat{R}_0^\theta, \ldots, \widehat{R}_{n-k}^\theta).$$

Now replace $D^j(\theta)$ by X_j $(j = 0, 1, 2, \ldots)$ on both sides of the equation. This completes the proof. $\Diamond$

Corollary 5.14. $(C_{n,k})$ and $(C_{n,k}^\perp)$ are B-representable.

Proof. Theorem 5.13 yields $C_{n,k} = A_{n,k} \circ \widehat{R}_{\sharp}$, whence by Eq. (2.14) and (2.8) $C_{n,k} = B_{n,k} \circ (A_{\sharp,1} \circ \widehat{R}_{\sharp})$, that is, $C_{n,k}$ is B-representable. — From Proposition 5.1 (iv) follows $C_{n,k}^{\perp} = B_{n,k}(\widehat{R}_0, \ldots, \widehat{R}_{n-k})$, hence $C_{n,k}^{\perp}$ is B-representable. ◇

Proposition 5.15.

(i) $C_{n,k}(1, \ldots, 1) = c(n, k)$

(ii) $C_{n,k}^{\perp}(1, \ldots, 1) = (-1)^{n-k} s_2(n, k)$

(iii) $C_{n,k}(1, 1, 0, \ldots, 0) = s_2(n, k)$

Proof. (i) Observing $\widehat{R}_j \circ 1 = P_{j,-1} \circ 1 = (-1)^j$ we obtain by Theorem 5.13 and Eq. (2.17)

$$C_{n,k}(1, \ldots, 1) = A_{n,k}(1, -1, \ldots, (-1)^{n-k})$$
$$= (-1)^{n-k} A_{n,k}(1, \ldots, 1)$$
$$= (-1)^{n-k} s_1(n, k) = c(n, k).$$

(ii) $C_{n,k}^{\perp}(1, \ldots, 1) = B_{n,k}(1, -1, 1, -1, \ldots) = (-1)^{n-k} B_{n.k}(1, \ldots, 1)$.
(iii) Set $\varphi := 1 + \iota$; then $C_{n,k}^{\varphi}(0) = C_{n,k}(1, 1, 0, \ldots, 0)$, and by Proposition 3.1 for every $j \geq 0$

$$\widehat{R}_j^{\varphi}(0) = D^j(\varphi^{-1})(0) = D^j((1 + \iota)^{-1})(0)$$
$$= (-1)^j j! = s_1(j + 1, 1) = A_{j+1,1}(1, \ldots, 1),$$

whence by Eq. (2.15) and Eq. (2.18)

$$C_{n,k}(1, 1, 0, \ldots, 0) = C_{n,k}^{\varphi}(0) = A_{n,k}(\widehat{R}_0^{\varphi}(0), \ldots, \widehat{R}_{n-k}^{\varphi}(0))$$
$$= A_{n,k}(A_{1,1}(1), \ldots, A_{n-k+1,1}(1, \ldots, 1))$$
$$= B_{n,k}(1, \ldots, 1) = s_2(n, k). ◇$$

Remark 5.9. The question of how the coefficients $\gamma_{n,k}(r_0, \ldots, r_{n-k})$ of (5.10) are made up in detail appears to be rather tricky and must remain open for the time being. Attempting a direct evaluation of $A_{n,k}(\widehat{R}_0, \ldots, \widehat{R}_{n-k})$ in general form quickly leads to a piling up of more and more cumbersome expressions. While the coefficients of $A_{n,k}$ and $B_{n,k}$ are products of simple combinatorial terms (see Eq. (2.13) and Eq. (2.16)), one might doubt whether this also applies to the coefficients of $C_{n,k}$. For example, consider

$$C_{n,n-4} = \binom{n}{5} \frac{15n^3 - 150n^2 + 485n - 502}{48} X_0^{n-4} X_1^4$$

$$+ \binom{n}{5} \frac{15n^2 - 85n + 116}{6} X_0^{n-3} X_1^2 X_2 + \binom{n}{5} \frac{5n - 13}{3} X_0^{n-2} X_2^2$$

$$+ \binom{n}{5} \frac{5n - 11}{2} X_0^{n-2} X_1 X_3 + \binom{n}{5} X_0^{n-1} X_4.$$

Here, the first two coefficients $\gamma_{n,n-4}(n-4,4)$ and $\gamma_{n,n-4}(n-3,2,1)$, regarded as polynomials in n, cannot be written as products of linear factors over the field $\mathcal{K}$. However, replacing each X_j by $\widehat{R}_j$ ($0 \le j \le 4$) and applying Corollary 4.2 (iii) resembles a magic wand that turns $C_{n,n-4} \circ \widehat{R}_\sharp$ into

$$A_{n,n-4}(X_0, \ldots, X_4) = 105 \binom{n+3}{8} X_0^{-n-4} X_1^4$$

$$- 105 \binom{n+2}{7} X_0^{-n-3} X_1^2 X_2 + 10 \binom{n+1}{6} X_0^{-n-2} X_2^2$$

$$+ 15 \binom{n+1}{6} X_0^{-n-2} X_1 X_3 - \binom{n}{5} X_0^{-n-1} X_4.$$

Also for the simpler cases $A_{n,n-k}$ with $k = 0, 1, 2, 3$ one gets similarly closed product representations of the coefficients as here (see Todorov [100, Equations (14) to (16)]). This could be seen as an indication that focusing on the iterations of θD ultimately turns out to be a less well-posed problem. The remedy is, of course, simply the choice $\theta = D(\varphi)^{-1}$ (see Eq. (2.6)).

6 Applications to binomial sequences

This section is to demonstrate the succinct way the classical topic of binomial sequences can be treated within the conceptual frame of the preceeding sections. Some new results will be proved.

6.1 Definition and representation

Let $f_n = f_n(t) \in \mathcal{K}[t]$ ($n = 0, 1, 2, 3 \ldots$) be a sequence of polynomials with deg $f_n = n$. Then, $f_0, f_1, f_2, \ldots$ is said to be *binomial*, or *of binomial type*, if for every $n \geq 0$

$$f_n(s + t) = \sum_{k=0}^{n} \binom{n}{k} f_{n-k}(s) f_k(t). \tag{6.1}$$

Note that clearly $f_0 = 1$ and $f'_n(0) \neq 0$.

The sequences t^n and $(t)_n$ are binomial (cf. Stanley [95, Exercise 5.37] for more examples). Knuth [49] also deals with binomial sequences f_n, but in their guise of *convolution polynomials* $f_n/n!$. As is well-known, binomial sequences are closely related to the exponential polynomials. This is reflected in the following two statements.

Proposition 6.1. *Let $f_0, f_1, f_2, \ldots$ be a sequence of polynomials from $\mathcal{K}[t]$. Then the following holds:*

(i) (f_n) *binomial* $\iff \exists \varphi \in \mathcal{G} : f_n(t) = [\frac{x^n}{n!}] e^{t\varphi(x)}$

(ii) (f_n) *binomial* $\iff f_n(t) = B_n(t f'_1(0), \ldots, t f'_n(0))$

Proof. (i) '$\Leftarrow$': Let φ be any invertible function and denote by $f_n(t)$ the nth Taylor coefficient of $e^{t\varphi}$; then it is easy to check that f_n satisfies the binomial property (6.1). — '$\Rightarrow$': Conversely, given f_n binomial, then $g(t, x) := \sum_{n \geq 0} f_n(t)(x^n/n!)$ satisfies $g(s + t, x) = g(s, x)g(t, x)$. Writing $g(t, x) = \sum_{n \geq 0} g_n(x)(t^n/n!)$, one gets $g_n(x) = g_1(x)^n$ for all $n \geq 0$ (by an inductive argument), and hence $g(t, x) = \sum_{n \geq 0}(g_1(x)t)^n/n! = e^{t g_1(x)}$. On the other hand we have $g_1(x) = \frac{\partial}{\partial t} g(t, x)|_{t=0} = \sum_{n \geq 1} f'_n(0)(x^n/n!)$ and therefore $g'_1(0) = f'_1(0) \neq 0$. Thus, taking $\varphi := g_1 \in \mathcal{G}$ delivers $[x^n/n!]e^{t\varphi} = [x^n/n!]g(t, x) = f_n(t)$.

(ii) According to (i) every binomial sequence can be written $f_n(t) = D^n(e^{t\varphi})(0)$. We have $\varphi(x) = \sum_{n \geq 1} f'_n(0)(x^n/n!)$ (as in the proof of (i)), whence $D^j(\varphi)(0) = f'_j(0)$. Now the FdB formula (2.12) yields $D^n(e^{t\varphi})(0) = \sum_{k=0}^{n} t^k B^{\varphi}_{n,k}(0) = B_n \circ t D^{\sharp}(\varphi)(0) = B_n(t f'_1(0), \ldots, t f'_n(0))$ (see also [84, Proposition 7.3 (i)]). $\diamond$

6.2 Some results on substitution

Several authors have dealt with evaluating the Bell polynomials in cases the variables have been replaced by special values from combinatorial number families and, more generally, also by binomial sequences (see, e.g., [22, 27,

67, 68, 102, 105]). Keeping in mind that, according to Proposition 6.1 (ii), the latter are instances of the exponential polynomials, it is not surprising that statements concerning the substitution of the B_n into polynomials can be readily transferred to the substitution of binomial sequences.

The subsequent list contains some of the more interesting results together with comments or short proofs. Suppose that (f_n) is a binomial sequence. Then the following statements hold:

$$L_n(f_1(t), \ldots, f_n(t)) = tf'_n(0), \tag{6.2}$$
$$P_{n,k}(f_1(t), \ldots, f_n(t)) = f_n(kt), \tag{6.3}$$
$$R_n(f_1(t), \ldots, f_n(t)) = f_n(-t), \tag{6.4}$$
$$T_n(f_1(t), \ldots, f_n(t)) = f_{n-1}(nt). \tag{6.5}$$

Proof. Eq. (6.2) follows directly from Proposition 6.1 (ii) by observing the inverse relationship between L_n and B_n (see Remark 4.1). Eq. (6.3) is derived from Proposition 4.5 by replacing each X_j with $tf'_j(0)$. Equations (6.4) and (6.5) are special cases of (6.3); recall that $R_n = P_{n,-1}$ and $T_n = P_{n-1,n}$. $\Diamond$

In the same way, without any calculation one gets from (3.11) and (6.3)

$$F_{n,k}(f_1(t), \ldots, f_n(t)) = \sum_{j=0}^{k} s_1(k,j) f_n(jt), \tag{6.6}$$

furthermore from Proposition 3.6

$$L_n(f_1(t), \ldots, f_n(t)) = \sum_{k=1}^{n} (-1)^{k-1} \frac{1}{k} \binom{n}{k} f_n(kt) \tag{6.7}$$

and likewise from Eq. (4.1)

$$B_{n,k}(f_1(t), \ldots, f_{n-k+1}(t)) = \frac{1}{k!} \sum_{j=0}^{k} (-1)^{k-j} \binom{k}{j} f_n(jt). \tag{6.8}$$

Remark 6.1. Eq. (6.8) has been proven by Yang [105, Theorem 2]. Mihoubi, by using the methods of ›Umbral Calculus‹, derived a slightly more general version [67, Proposition 2]. Equations (6.3)–(6.7) are new.

From Eq. (5.5) one immediately obtains by (6.3)

$$B_{n,k}(f_0(t), 2f_1(t), 3f_2(t), \ldots) = \binom{n}{k} f_{n-k}(kt). \qquad (6.9)$$

Then, Proposition 5.7 yields the corresponding orthogonal companion

$$A_{n,k}(f_0(t), 2f_1(t), 3f_2(t), \ldots) = \binom{n-1}{k-1} f_{n-k}(-nt). \qquad (6.10)$$

Furthermore, by substituting (6.5) into the equations of Theorem 5.6 we readily get

$$B_{n,k}(f_0(t), f_1(2t), f_2(3t), \ldots) = \binom{n-1}{k-1} f_{n-k}(nt). \qquad (6.11)$$

together with the orthogonal companion

$$A_{n,k}(f_0(t), f_1(2t), f_2(3t), \ldots) = \binom{n}{k} f_{n-k}(-kt). \qquad (6.12)$$

Remark 6.2. Eq. (6.9) has been proved by Yang [105, Theorem 1]. Abbas and Bouroubi [1, Theorem 3] have shown (6.11) for the special case $t = 1$, which just represents the binomial variant of the Khelifa/Cherruault identity mentioned in Remark 5.6. The identities (6.10) and (6.12) are new.

6.3 A binomial sequence related to trees

Let τ denote the exponential generating function for labeled rooted trees, that is, $\tau(x) = \sum_{n\geq 1} n^{n-1} x^n / n!$ (see Remark 3.3 (iii)). Knuth and Pittel [52] have introduced univariate expressions $t_n(y)$, $n \geq 0$, called 'tree polynomials' and defined by

$$t_n(y) := \left[\frac{x^n}{n!}\right] \frac{1}{(1 - \tau(x))^y}.$$

Defining the functions $g, \varphi \in \mathcal{G}$ by $g(x) := x(1 - x)^{-1}$ and $\varphi := \lambda \circ g \circ \tau$, we obtain $(1 - \tau(x))^{-y} = e^{y\varphi(x)}$. From this it follows by Proposition 6.1 (i) that

$$t_0(y) = 1, \quad t_1(y) = y, \quad t_2(y) = 3y + y^2, \quad t_3(y) = 17y + 9y^2 + y^3, \ldots$$

is actually a binomial sequence of polynomials from $\mathbb{Z}[y]$. If we write $t_n(y) = \sum_{k=0}^{n} t_{n,k} y^k$, then the coefficient $t_{n,k}$ counts the total number of mappings of $\{1, 2, \ldots, n\}$ into itself having exactly k different cycles.

Knuth and Pittel derived some integral formulas and a Γ-function representation of the $t_n(y)$ (see ibid., equations (2.12), (2.13)), inasmuch the focus of their paper is on the asymptotic behavior of these polynomials. Supplementary to this, we will prove here a new explicit representation built up exclusively by means of elementary combinatorial operations.

Theorem 6.2.

(i) $\qquad t_{n,k} = \sum_{k \leq j \leq i \leq n} s_1(j,k) \frac{n^{n-i} i!}{j!} \binom{i-1}{j-1} \binom{n-1}{i-1};$

(ii) $\qquad t_n(y) = B_n(yt_{1,1}, \ldots, yt_{n,1}),$ *where for* $r \geq 1$

$$t_{r,1} = \sum_{1 \leq j \leq i \leq r} (-1)^{j-1} \frac{r^{r-i} i!}{j} \binom{i-1}{j-1} \binom{r-1}{i-1}.$$

Proof. (i) With the terms introduced above we have

$$\sum_{n \geq 0} t_n(y) \frac{x^n}{n!} = e^{y \varphi(x)} = \sum_{n \geq 0} \left(\sum_{k=0}^{n} B_{n,k}^{\varphi}(0) y^k \right) \frac{x^n}{n!}$$

(cf. [84, Proposition 7.3 (i)]), whence $t_n(y) = \sum_{k=0}^{n} B_{n,k}^{\varphi}(0) y^k$ and by Jabotinsky's Theorem 4.7 (i)

$$t_{n,k} = B_{n,k}^{\varphi}(0) = B_{n,k}^{\lambda \circ (g \circ \tau)}(0) = \sum_{j=k}^{n} B_{n,j}^{g \circ \tau}(0) B_{j,k}^{\lambda}(0). \qquad (*)$$

Now observe that $B_{n,k}^{\lambda}(0) = B_{n,k}^{\bar{\varepsilon}}(0) = A_{n,k}^{\varepsilon}(0) = s_1(n,k)$; thus again applying Jabotinsky's formula, we obtain from (*)

$$t_{n,k} = \sum_{j=k}^{n} s_1(j,k) \sum_{i=j}^{n} B_{n,i}^{\tau}(0) B_{i,j}^{g}(0). \qquad (**)$$

Evaluating both substitutions gives

$$B_{n,i}^{\tau}(0) = B_{n,i}(1^0, 2^1, 3^2, \ldots) = \binom{n-1}{i-1} n^{n-i}$$

by applying Theorem 5.6 (i) (forest numbers, cf. Remark 5.6), and

$$B_{i,j}^{g}(0) = B_{i,j}(1!, 2!, 3!, \ldots) = l^{+}(i,j) = \frac{i!}{j!}\binom{i-1}{j-1}$$

(unsigned Lah numbers, cf. (5.8)). Finally, we substitute these results into (**) thus arriving at the asserted statement.

(ii) Recall that $s_1(j, 1) = (-1)^{j-1}(j-1)!$. Thus $t_{r,1}$ is obtained from (i) by taking $n = r$ and $k = 1$. Furthermore $t_r'(0) = t_{r,1}$, whence the assertion follows according to Proposition 6.1 (ii). ◇

6.4 Coupling binomial sequences

In their seminal paper [72] Mullin and Rota developed a general theory of binomial sequences. They introduced a family of shift-invariant linear differential operators (called 'delta operators') on the vector space of polynomials such that each sequence of binomial type can be associated uniquely to a specific ('basic') delta operator. On that basis, they were able to describe the exact form of the connection between two given binomial sequences (see also Aigner [3, Chapter III]).

In the following it will be demonstrated how the main result of these investigations can also (alternatively) be established by using only properties of the Bell polynomials.

Theorem 6.3 (Mullin & Rota 1970). *Let (f_n) and (g_n) be any two binomial sequences. Then there exist unique constants $c_{n,k} \in K$, with $c_{n,k} = 0$ for $n < k$, such that $f_n(t) = \sum_{k=0}^{n} c_{n,k}g_k(t)$ for every $n \geq 0$, and the sequence (h_n) defined by $h_n(t) := \sum_{k=0}^{n} c_{n,k}t^k$ is of binomial type.*

Proof. According to Proposition 6.1 there are functions $\varphi, \psi \in \mathcal{G}$ such that f_n and g_n can be written in the form

$$g_n(t) = \sum_{k=0}^{n} t^k B_{n,k}^{\varphi}(0) \quad \text{and} \quad f_n(t) = \sum_{k=0}^{n} t^k B_{n,k}^{\psi}(0). \qquad (6.13)$$

Since every sequence of binomial type consists of polynomials linearly independent in $\mathcal{K}[t]$, it follows that

$$f_n(t) = \sum_{k=0}^{n} c_{n,k}g_k(t) \qquad (6.14)$$

with uniquely determined connecting coefficients $c_{n,k} \in \mathcal{K}$. Now substituting the right-hand sides of (6.13) into (6.14) we obtain

$$\sum_{k=0}^{n} t^k B_{n,k}^{\psi}(0) = \sum_{k=0}^{n} c_{n,k} \sum_{j=0}^{k} t^j B_{k,j}^{\varphi}(0)$$

$$= \sum_{k=0}^{n} t^k \sum_{j=k}^{n} c_{n,j} B_{j,k}^{\varphi}(0)$$

after rearranging the double series. Hence, by equating the coefficients of t^k

$$B_{n,k}^{\psi}(0) = \sum_{j=k}^{n} c_{n,j} B_{j,k}^{\varphi}(0). \tag{6.15}$$

On the other hand, applying Jabotinsky's formula (Theorem 4.7 (i)) to the 'decomposition' $\psi = \varphi \circ (\overline{\varphi} \circ \psi)$ yields

$$B_{n,k}^{\psi}(0) = \sum_{j=k}^{n} B_{n,j}^{\overline{\varphi} \circ \psi}(0) B_{j,k}^{\varphi}(0). \tag{6.16}$$

We now set $V_{n,j} := c_{n,j} - B_{n,j}^{\overline{\varphi} \circ \psi}(0)$. Then, from (6.15) and (6.16)

$$\sum_{j=k}^{n} V_{n,j} B_{j,k}(\varphi_1, \ldots, \varphi_{j-k+1}) = 0 \qquad (0 \le k \le n), \tag{6.17}$$

where $\varphi_1, \varphi_2, \varphi_3, \ldots$ are the Taylor coefficients of φ. We take for k the values $n, n-1, n-2, \ldots$ in that order to show $V_{n,j} = 0$ for $j = n, n-1, \ldots, 0$. In the case $k = n$ (6.17) becomes $V_{n,n} B_{n,n}(\varphi_1) = V_{n,n} \varphi_1^n = 0$, hence $V_{n,n} = 0$ because of $\varphi_1 \ne 0$. For $k = n-1$ we similarly get $0 = V_{n,n-1} B_{n-1,n-1}(\varphi_1) + V_{n,n} B_{n,n-1}(\varphi_1) = V_{n,n-1} \varphi_1^{n-1}$, that is, $V_{n,n-1} = 0$, and so on until finally $V_{n,0} = 0$. All in all, (6.17) implies

$$c_{n,j} = B_{n,j}^{\overline{\varphi} \circ \psi}(0) = B_{n,j}(a_1, \ldots, a_{n-j+1}) \qquad (0 \le j \le n)$$

with $a_j = D^j(\overline{\varphi} \circ \psi)(0)$. Clearly we now have $c_{n,j} = 0$ for $n < j$, and furthermore

$$h_n(t) = \sum_{k=0}^{n} c_{n,k} t^k = \sum_{k=0}^{n} t^k B_{n,k}(a_1, \ldots, a_{n-k+1}) = B_n(ta_1, \ldots, ta_n),$$

where $h'_j(0) = a_j$. Hence (h_n) is binomial by Proposition 6.1 (ii).   ◊

From the above proof we learn that for any two given binomial sequences their connecting coefficients are B-representable. The question here is whether one of the sequences can be of binomial type while the other is not. The (negative) answer is provided by the following *both-or-none statement*.

Theorem 6.4. *Let (f_n) and (g_n) be any sequences of polynomials from $\mathcal{K}[t]$ and $a_1\,(\neq 0), a_2, a_3, \ldots$ a sequence of constants from $\mathcal{K}$. Suppose that*

$$f_n(t) = \sum_{k=0}^{n} g_k(t) B_{n,k}(a_1, \ldots, a_{n-k+1})$$

holds for all $n \geq 0$. Then we have: g_n binomial $\iff$ f_n binomial .

Proof. 1. Assume $g_n(t)$ to be of binomial type; then, as in the proof of Theorem 6.3, $g_n(t) = \sum_{k=0}^{n} t^k B_{n,k}^{\varphi}(0)$ with $\varphi \in \mathcal{G}$. We set $\psi(x) := \sum_{n \geq 1} a_n(x^n/n!)$, which turns out to be invertible because of $\psi'(0) = a_1 \neq 0$. From this we obtain

$$f_n(t) = \sum_{k=0}^{n} \left(\sum_{j=0}^{k} t^j B_{k,j}^{\varphi}(0) \right) B_{n,k}^{\psi}(0)$$

$$= \sum_{k=0}^{n} t^k \left(\sum_{j=k}^{n} B_{n,j}^{\psi}(0) B_{j,k}^{\varphi}(0) \right) \quad \text{(by rearranging the double series)}$$

$$= \sum_{k=0}^{n} t^k B_{n,k}^{\varphi \circ \psi}(0). \qquad\qquad \text{(by Jabotinsky's formula)}$$

It follows by Proposition 6.1 that $f_n(t)$ is of binomial type.

2. Conversely, suppose $f_n(t)$ is binomial. This case will be reduced to the situation of part 1. With the same denotations we have $f_n(t) = \sum_{k=0}^{n} g_k(t) B_{n,k}^{\psi}(0)$, $\psi \in \mathcal{G}$. To this now, the generalized Stirling inversion (Proposition 5.4) can be applied thus yielding

$$g_n(t) = \sum_{k=0}^{n} f_k(t) A_{n,k}^{\psi}(0)$$

$$= \sum_{k=0}^{n} f_k(t) B_{n,k}^{\overline{\psi}}(0) = \sum_{k=0}^{n} f_k(t) B_{n,k}(\overline{a}_1, \ldots, \overline{a}_{n-k+1}),$$

where $\bar{a}_j = D^j(\bar{\psi})(0)$. Since $\bar{a}_1 = 1/a_1 \neq 0$, we see from the statement of part 1 that (g_n) is binomial. $\diamond$

With any binomial sequence it is now easy to create a new one by linearly combining its polynomials with instances of the partial Bell polynomials as connecting coefficients. The following special case has been established by Yang [105, Lemma 2, p. 53].

Corollary 6.5. *A sequence (f_n) is binomial, if and only if for every $n \geq 0$*

$$f_n(t) = \sum_{k=0}^{n} (t)_k B_{n,k}(f_1(1), \ldots, f_{n-k+1}(1)).$$

Proof. Since we have

$$(t)_n = \sum_{k=0}^{n} t^k s_1(n,k) = \sum_{k=0}^{n} t^k B_{n,k}^{\lambda}(0) = \left[\frac{x^n}{n!}\right] e^{t\lambda(x)},$$

$(t)_n$ turns out to be of binomial type (by Proposition 6.1 (i)). Suppose now

$$f_n(t) = \sum_{k=0}^{n} (t)_k B_{n,k}(a_1, \ldots, a_{n-k+1}) \tag{*}$$

with $a_1 \neq 0$. One clearly has $f_j(1) = a_j$, and $f_n(t)$ is binomial by Theorem 6.4. — Conversely, assume $f_n(t)$ is binomial. The Mullin-Rota Theorem 6.3 then yields connecting coefficients $B_{n,k}(a_1, \ldots, a_{n-k+1})$ such that equation (*) is satisfied. $\diamond$

7 Lagrange inversion polynomials

7.1 Some preliminaries

After having encountered various inverse relationships based on orthogonality in previous sections, we now turn to the type of *compositional inversion* already mentioned in Example 4.1 and Remark 4.3. First we want to free ourselves from the assumption made there, according to which the polynomials in question must be FdB, i. e., elements of $\Phi_n[\mathcal{G}]$. Therefore, we will consider sequences of arbitrary polynomials (U_n) and (V_n) such that for every $n \geq 0$

$$U_n \circ V_\sharp = X_n \quad \text{and} \quad V_n \circ U_\sharp = X_n. \tag{7.1}$$

However, the idea that a polynomial sequence remains (uniquely) associated to a function should still be adhered to. In order to roughly sketch the intended situation here, let us assume that $f, g \in \mathcal{F}$ are somehow characterized by sequences of constants $c_0, c_1, c_2, \ldots$ and $d_0, d_1, d_2, \ldots$, respectively. Next suppose there is a sequence of polynomials (U_n) such that $d_n = U_n(c_0, \ldots, c_n)$ for every $n \geq 0$. Then U_n will be called *conversion polynomial* (of f with respect to g). Reversely, if we additionally have conversion polynomials V_n of g (w.r.t. f), then the pair (U_n, V_n) satisfies (7.1). In the special case $g = \overline{f}$, the corresponding conversion polynomials will henceforth be called *(generalized) Lagrange inversion polynomials*.

The classical Lagrange inversion is a special case that arises when f and $\overline{f}$ are characterized by their respective Taylor coefficients $f_1, f_2, \ldots$ and $\overline{f}_1, \overline{f}_2, \ldots$. While a major part of the extensive literature on the subject is devoted to its analytical and combinatorial aspects (e. g. [54, 30]), explicit formulas for the coefficients of $\overline{f}$ in the form of polynomial expressions

$$\overline{f}_n = \Lambda_n(f_1, \ldots, f_n)$$

have been studied at times, for instance [104], [71, p. 412], [30, Section 2.5]. Comtet [22, p. 151] replaced these determinantal and diophantine representations with the following elegant formula (the prehistory of which is sketched in [43]):

$$\Lambda_n = \sum_{k=0}^{n-1} (-1)^k X_1^{-(n+k)} B_{n-1+k,k}(0, X_2, \ldots, X_n). \tag{7.2}$$

Eq. (7.2) is a special case of Theorem 2.3 ($=$ Theorem 6.1 in [84]), from which immediately follows that $\Lambda_n = A_{n,1}$. Alternatively, if one wants to avoid the associate Bell polynomials $\widetilde{B}_{n-1+k,k}$ in (7.2), the Schlömilch-Schläfli representation (5.2) could be used (taking $k = n - 1$) to obtain

$$\Lambda_n = \sum_{k=0}^{n-1} (-1)^k \binom{2n-1}{n-1-k} X_1^{-(n+k)} B_{n-1+k,k}.$$

Finally, according to Corollary 3.4, Λ_n can also be expressed by virtue of the reciprocal polynomials and the tree polynomials.

Since the Lagrange inversion polynomial Λ_n is the conversion polynomial of f (w.r.t. $\overline{f}$) and of $\overline{f}$ (w.r.t. f) as well, it obviously must be self-inverse: $\Lambda_n \circ \Lambda_\sharp = X_n$ (see also Eq. (2.15) for $k = 1$).

In addition, we see from Theorems 3.7 and 3.9 that $B_{n,k}(f_1, \ldots, f_{n-k+1})$ and $A_{n,k}(f_1, \ldots, f_{n-k+1})$ are the Taylor coefficients of f^k and $\overline{f}^k$, respectively. Again we encounter an orthogonality relation, this time through raising a function and its inverse to the kth power.

7.2 A modified inversion problem

We start with a theorem that gives the solution to a modified inversion problem.

Theorem 7.1. *Let s be any positive integer and let $c_0, c_1, c_2, \ldots \in \mathcal{K}$ be any sequence of constants with $c_0 = 1$. Then, $f(x) = \sum_{n \geq 0} \frac{c_n}{n!} x^{sn+1}$ has an inverse of the form $\overline{f}(x) = \sum_{n \geq 0} \frac{d_n}{n!} x^{sn+1}$ with $d_0 = 1$ and $d_n = \Lambda_n(c_1, \ldots, c_n)$ for every $n \geq 1$, where*

$$\Lambda_n = \sum_{k=1}^{n} (-1)^k \binom{sn+k}{k-1} (k-1)! B_{n,k}.$$

This statement is Mihoubi's [68] slightly modified (symmetrized) version of a theorem established by Comtet [22, Theorem F, p. 151]. The possibly first proof to be found in the literature results from specializing a pair of inverse polynomials investigated by Birmajer, Gil and Weiner [10, Theorem 4.6, Example 4.9]. In the following, too, we will regain Comtet's theorem as a corollary of a much more general statement. However, to achieve this, we keep following the idea (outlined in Section 7.1) of switching between certain representations of a function and its inverse using generalized Lagrange inversion polynomials. The one from Theorem 7.1, Λ_n, is the conversion polynomial of f (characterized by $c_0, c_1, c_2, \ldots$) w.r.t. $\overline{f}$ (characterized by $d_0, d_1, d_2, \ldots$), and *vice versa*, that is, we obviously have $\Lambda_n \circ \Lambda_\sharp = X_n$.

In order to obtain a more comprehensive class of inversion polynomials, we will modify the assumptions made in Theorem 7.1 as follows: Suppose f is any invertible function and denote by f_n and $\overline{f}_n$, $n \geq 1$, the respective Taylor coefficients of f and $\overline{f}$. Both functions are now to be given the new form

$$\text{(i)} \ \ f(x) = \sum_{n \geq 0} \frac{c_n}{n!} a(x) \varphi(x)^n \quad \text{and} \quad \text{(ii)} \ \ \overline{f}(x) = \sum_{n \geq 0} \frac{d_n}{n!} b(x) \psi(x)^n, \quad (7.3)$$

where a, b and φ, ψ are assumed to be fixed functions (with suitable properties). We will see that the required conversions can be achieved by merely adjusting the sequences $c_0, c_1, c_2, \ldots$ and $d_0, d_1, d_2, \ldots$. In a final step, Lagrange inversion polynomials U_n, V_n with property (7.1) will be constructed satisfying $d_n = U_n(c_0, \ldots, c_n)$ and $c_n = V_n(d_0, \ldots, d_n)$.

First we have to fix the conditions that guarantee the existence of the above representations (7.3). For this purpose we define the functions

$$c(x) = \sum_{n \geq 0} c_n \frac{x^n}{n!} \quad \text{and} \quad d(x) = \sum_{n \geq 0} d_n \frac{x^n}{n!}$$

so that the series (7.3) can equivalently be rewritten as function terms:

$$\text{(i)} \ f = a \cdot (c \circ \varphi) \quad \text{and} \quad \text{(ii)} \ \overline{f} = b \cdot (d \circ \psi). \tag{7.4}$$

It is at once clear that f, φ, ψ must be invertible functions (what will be tacidly assumed for the remainder of this section). Furthermore, we generally suppose that both c and f are neither part of a nor of φ, and that the corresponding equally holds for d and $\overline{f}$ with respect to b and ψ. As usual, the Taylor coefficients of a and b are denoted by $a_0, a_1, a_2, \ldots$ and $b_0, b_1, b_2, \ldots$, respectively.

Let us first turn to the equation (i) in (7.4).

Proposition 7.2. *Suppose that equation* $(7.4, \text{i})$ *holds. Then, for all* $n \geq 0$: $f_n = \Gamma_n(a, \varphi)(c_0, \ldots, c_n)$, *where*

$$\Gamma_n(a, \varphi) := \sum_{k=0}^{n} \left(\sum_{j=k}^{n} \binom{n}{j} a_{n-j} B_{j,k}^{\varphi}(0) \right) X_k.$$

Proof. According to Definition 3.1 the conversion polynomial $\Gamma_n(a, \varphi)$ we are looking for is $\Omega_n(f \mid c)$. Therefore, we obtain by Eq. (3.5)

$$\Gamma_n(a, \varphi) = \Omega_n(a \cdot (c \circ \varphi) \mid c) = \sum_{j=0}^{n} \binom{n}{j} \Omega_{n-j}(a \mid c) \Omega_j(c \circ \varphi \mid c). \tag{*}$$

Since c does not appear in either a or φ, we have $\Omega_{n-j}(a \mid c) = a_{n-j}$ and observing (3.6) and (3.3)

$$\Omega_j(c \circ \varphi \mid c) = \sum_{k=0}^{j} \Omega_k(c \mid c)(B_{j,k} \circ \Omega_\sharp(\varphi \mid c)) = \sum_{k=0}^{j} B_{j,k}^{\varphi}(0) X_k.$$

Substitution into (*) and rearranging the double series in $\Gamma_n(a, \varphi)$ yields the assertion. ◊

Remark 7.1. Note that the requirements for Proposition 7.2 can be weakend to $f \in \mathcal{F}$ and $\varphi \in \mathcal{F}_0$. Under these assumptions, the special case $a = 1$ is of interest here. It yields

$$\sum_{n \geq 0} c_n \frac{\varphi(x)^n}{n!} = \sum_{n \geq 0} \left(\sum_{k=0}^{n} B_{n,k}^{\varphi}(0) c_k \right) \frac{x^n}{n!},$$

which, by further specializing $c_k = t^k$, leads to the expansion of $e^{t\varphi(x)}$ used in Section 6 (Proposition 6.1); see also [84, Proposition 7.3].

Applying Remark 2.1 to (7.4, i) we see that f is invertible in just the following two cases, both of which are covered by the linear form $\Gamma_n(a, \varphi)$:

(1) $a \in \mathcal{F}_1$ and $c \circ \varphi \in \mathcal{G}$,

(2) $a \in \mathcal{G}$ and $c \circ \varphi \in \mathcal{F}_1$.

The inverse transformations that compute the c_n's from the f_n's also turn out to be linear forms. There are two case-specific types that differ from one another.

Proposition 7.3. *Suppose that* (7.4, i) *and* $a \in \mathcal{F}_1$. *Then, for every* $n \geq 0$:
$c_n = \overline{\Gamma}_n^{(1)}(a, \varphi)(0, f_1, \ldots, f_n)$, *where*

$$\overline{\Gamma}_n^{(1)}(a, \varphi) := \sum_{k=0}^{n} \left(\sum_{j=k}^{n} \binom{j}{k} \widehat{R}_{j-k}(a_0, \ldots, a_{j-k}) A_{n,j}^{\varphi}(0) \right) X_k.$$

Proof. In order to obtain c_n, we transform (7.4, i) into

$$c = (a^{-1} \cdot f) \circ \overline{\varphi}. \tag{7.5}$$

From this we get the inverse form

$$\overline{\Gamma}_n^{(1)}(a, \varphi) = \Omega_n(c \,|\, f) = \Omega_n((a^{-1} \cdot f) \circ \overline{\varphi} \,|\, f)$$

$$= \sum_{j=0}^{n} \Omega_j(a^{-1} \cdot f \,|\, f)(B_{n,j} \circ \Omega_\sharp(\overline{\varphi} \,|\, f)) \tag{by (3.6)}$$

$$= \sum_{j=0}^{n} \sum_{k=0}^{j} \binom{j}{k} \Omega_{j-k}(a^{-1} \,|\, f) \Omega_k(f \,|\, f) B_{n,j}^{\overline{\varphi}}(0) \tag{by (3.5)}$$

$$= \sum_{j=0}^{n} \sum_{k=0}^{j} \binom{j}{k} \widehat{R}_{j-k}(a_0, \dots, a_{j-k}) A_{n,j}^{\varphi}(0) X_k. \text{ (by Prop. 3.1, (2.11))}$$

The assertion now follows by rearranging the double series as in the proof of Proposition 7.2.   ◊

Proposition 7.4. *Suppose that* (7.4, i) *and* $a \in \mathcal{G}$. *Then, for every* $n \geq 0$:
$c_n = \overline{\Gamma}_n^{(2)}(a, \varphi)(f_1, \dots, f_{n+1})$, *where*

$$\overline{\Gamma}_n^{(2)}(a, \varphi) := \sum_{k=0}^{n} \frac{1}{k+1} \left(\sum_{j=k}^{n} \binom{j}{k} \widehat{R}_{j-k}(\tfrac{a_1}{1}, \dots, \tfrac{a_{j-k+1}}{j-k+1}) A_{n,j}^{\varphi}(0) \right) X_{k+1}.$$

Proof. Again we use Eq. (7.5), but now with $a_0 = 0$ and $a_1 \neq 0$. Defining $h \in \mathcal{F}_1$ by $h(x) := \sum_{n \geq 0} \frac{a_{n+1}}{n+1} \frac{x^n}{n!}$ we obtain $a(x) = x \cdot h(x)$, and hence $a^{-1} \cdot f = h^{-1} \cdot (f/\iota)$. As in the proof of Proposition 7.3, it follows

$$\overline{\Gamma}_n^{(2)}(a, \varphi) = \Omega_n(c \,|\, f) = \Omega_n((h^{-1} \cdot \tfrac{f}{\iota}) \circ \overline{\varphi} \,|\, f)$$

$$= \sum_{j=0}^{n} \Omega_j(h^{-1} \cdot \tfrac{f}{\iota} \,|\, f) A_{n,j}^{\varphi}(0).$$

Now we have by (3.5) and Proposition 3.1

$$\Omega_j(h^{-1} \cdot \tfrac{f}{\iota} \,|\, f) = \sum_{k=0}^{j} \binom{j}{k} \widehat{R}_{j-k}(\tfrac{a_1}{1}, \dots, \tfrac{a_{j-k+1}}{j-k+1}) \Omega_k(\tfrac{f}{\iota} \,|\, f).$$

Observing $\Omega_k(\tfrac{f}{\iota} \,|\, f) = \frac{1}{k+1} X_{k+1}$ and rearranging the double series completes the proof.   ◊

For the sake of clarity we will distinguish the cases (1) $a \in \mathcal{F}_1$ and (2) $a \in \mathcal{G}$ also for $\Gamma_n(a, \varphi)$ in Proposition 7.2. We set $\Gamma_n^{(1)}(a, \varphi) := \Gamma_n(a, \varphi)(0, X_1, \dots, X_n)$, since $a_0 \neq 0$ requires $c_0 = 0$ in (7.4, i). Otherwise, we use $\Gamma_n^{(2)}(a, \varphi)$ to denote the result obtained by putting $a_0 = 0$ in $\Gamma_n(a, \varphi)$.

Now we get from Propositions 7.2, 7.3, 7.4 the

Corollary 7.5. *For every* $n \geq 1$ *we have*

$$\Gamma_n^{(i)}(a, \varphi) \circ \overline{\Gamma}_{\sharp}^{(i)}(a, \varphi) = \overline{\Gamma}_n^{(i)}(a, \varphi) \circ \Gamma_{\sharp}^{(i)}(a, \varphi) = X_n \quad (i = 1, 2).$$

Convention. Similarly as $\Gamma_n(a, \varphi)$ covers both cases (1) $a \in \mathcal{F}_1$ and (2) $a \in \mathcal{G}$, we formally write for brevity $\overline{\Gamma}_n(a, \varphi)$, which is to mean either $\overline{\Gamma}_n^{(1)}(a, \varphi)$ or $\overline{\Gamma}_n^{(2)}(a, \varphi)$ depending on which of the two cases actually occurs.

We are now in a position to construct the desired general type of Lagrange inversion polynomial that fits the situation given in Eq. (7.3).

Definition 7.1 (Generalized Lagrange inversion polynomial).
Given any $\varphi, \psi \in \mathcal{G}$ and $a, b \in \mathcal{F}_1 \cup \mathcal{G}$, we define

$$\Lambda_n(a, \varphi \,|\, b, \psi) := \overline{\Gamma}_n(b, \psi) \circ A_{\sharp,1} \circ \Gamma_\sharp(a, \varphi).$$

Theorem 7.6. *Suppose* $f, \varphi, \psi \in \mathcal{G}$ *and* $a, b \in \mathcal{F}_1 \cup \mathcal{G}$. *Then, there are uniquely determined sequences of constants* (c_n) *and* (d_n) *such that the equations* (7.3) *hold. Moreover, we have* $d_n = \Lambda_n(a, \varphi \,|\, b, \psi)(c_0, \ldots, c_n)$ *and* $c_n = \Lambda_n(b, \psi \,|\, a, \varphi)(d_0, \ldots, d_n)$, *or equivalently*

$$\Lambda_n(a, \varphi \,|\, b, \psi) \circ \Lambda_\sharp(b, \psi \,|\, a, \varphi) = X_n \qquad \text{for every } n \geq 1.$$

Proof. From f we obtain (c_n) by applying $\overline{\Gamma}_n(a, \varphi)$ according to Propositions 7.3 and 7.4; and in the same unique way one gets (d_n) by applying $\overline{\Gamma}_n(b, \psi)$ to the Taylor coefficients of $\overline{f}$.

Upon closer inspection, it becomes clear that in general we have $\Lambda_n(a, \varphi \,|\, b, \psi) \in \mathcal{K}[X_0^{-1}, X_1^{-1}, X_0, X_1, \ldots, X_{n+1}]$. Depending on the choice of a, b the (Laurent) polynomial $\Lambda_n(a, \varphi \,|\, b, \psi)$ can take four different forms. It may suffice here to consider, for example, the case $a \in \mathcal{G}$ and $b \in \mathcal{F}_1$. Observing $f_0 = \overline{f}_0 = 0$ we obtain

$$
\begin{aligned}
d_n &= \overline{\Gamma}_n^{(1)}(b, \psi)(0, \overline{f}_1, \ldots, \overline{f}_n) && \text{(Proposition 7.3)} \\
&= \overline{\Gamma}_n^{(1)}(b, \psi)(A_{0,1}(0), A_{1,1}(f_1), \ldots, A_{n,1}(f_1, \ldots, f_n)) && \text{(Remark 2.4)} \\
&= \overline{\Gamma}_n^{(1)}(b, \psi) \circ A_{\sharp,1} \circ f_\sharp \\
&= \overline{\Gamma}_n^{(1)}(b, \psi) \circ A_{\sharp,1} \circ \Gamma_\sharp^{(2)}(a, \varphi)(c_0, \ldots, c_{\sharp-1}, 0) && \text{(Proposition 7.2)} \\
&= \Lambda_n(a, \varphi \,|\, b, \psi)(c_0, \ldots, c_{n-1}, 0). && \text{(Definition 7.1)}
\end{aligned}
$$

In the last step, the indeterminate X_j in $\overline{\Gamma}_n^{(1)}(b, \psi)(0, A_{1,1}, \ldots, A_{n,1})$ is replaced by $\Gamma_j^{(2)}(a, \varphi)(c_0, \ldots, c_{j-1}, 0)$ for each $j = 1, \ldots, n.$ — In the reverse direction one obtains in a similar way

$$c_n = \overline{\Gamma}_n^{(2)}(a, \varphi)(f_1, \ldots, f_{n+1}) \qquad\qquad \text{(Proposition 7.4)}$$

$$= \overline{\Gamma}_n^{(2)}(a,\varphi) \circ A_{\sharp,1} \circ \overline{f}_\sharp$$

$$= \overline{\Gamma}_n^{(2)}(a,\varphi) \circ A_{\sharp,1} \circ \Gamma_\sharp^{(1)}(b,\psi)(0,d_1,\ldots,d_\sharp)$$

$$= \Lambda_n(b,\psi \,|\, a,\varphi)(0,d_1,\ldots,d_n). \qquad\qquad \Diamond$$

In the general case, the inversion polynomials $\Lambda_n(a,\varphi\,|\,b,\psi)$ turn out to be rather complicated expressions even for small values of n. We therefore confine ourselves here to the study of two interesting special cases: $a = b = 1$ and $a = b = \iota$.

Theorem 7.7. *Let φ,ψ be invertible functions. Then, for every $n \geq 1$*

$$\Lambda_n(1,\varphi \,|\, 1,\psi) = \sum_{k=1}^{n} \left(A_{n,k}^\psi(0) \sum_{j=1}^{k} A_{j,1}^\varphi(0) A_{k,j} \right).$$

Proof. Since $b_0 = 1$ and $b_n = 0$, $n \geq 1$, we have $\widehat{R}_{j-k}(b_0,b_1,\ldots,b_{j-k}) = \delta_{jk}$, and hence $\overline{\Gamma}_n^{(1)}(1,\psi) = \sum_{k=1}^{n} A_{n,k}^\psi(0) X_k$ by Proposition 7.3. Thus it follows from Theorem 7.6

$$\Lambda_n(1,\varphi \,|\, 1,\psi) = \sum_{k=1}^{n} A_{n,k}^\psi(0) A_{k,1}(\Gamma_1(1,\varphi),\ldots,\Gamma_k(1,\varphi)). \qquad (7.6)$$

Now consider an arbitrary $f \in \mathcal{G}$ of the form $f = c \circ \varphi$. Then $c_0 = 0$, $c_1 \neq 0$ and $f_0 = 0$, $f_1 \neq 0$, and by Proposition 7.2: $f_n = \Gamma_n(1,\varphi)(0,c_1,\ldots,c_n)$ for $n \geq 0$. Applying (7.6) to $(c_0,\ldots,c_n)$ thus yields

$$\Lambda_n(1,\varphi \,|\, 1,\psi)(0,c_1,\ldots,c_n) = \sum_{k=1}^{n} A_{n,k}^\psi(0) A_{k,1}^f(0). \qquad (7.7)$$

Finally, we obtain by Theorem 4.7 (ii)

$$A_{k,1}^f(0) = \sum_{j=k} A_{j,1}^\varphi(0) A_{k,j}(c_1,\ldots,c_{k-j+1}).$$

Substitution of the latter into (7.7) gives the assertion. $\qquad\qquad \Diamond$

Remark 7.2. $\Lambda_n(1,\varphi\,|\,1,\psi)$ solves the problem of inverting a function $c \circ \varphi \in \mathcal{G}$ into a function of the prescribed form $d \circ \psi$ by computing the sequence $d_1,d_2,\ldots$ from $c_1,c_2,\ldots$. In the particular case $\varphi = \psi = \iota$ this amounts to the classical Lagrange inversion. Indeed, observing $A_{n,k}^\iota(0) = A_{n,k}(1,0,\ldots,0) = \delta_{nk}$ we get from Theorem 7.7: $\Lambda_n(1,\iota\,|\,1,\iota) = A_{n,1}$.

Our second case, $a = b = \iota$, is a little more involved; its solution at least can be formulated more succinctly by introducing the following special polynomial:

$$\widehat{I}_{n,k} := \sum_{j=0}^{n} (-1)^j (k+j)_{j-1} X_0^{-(k+1+j)} B_{n,j}.$$

We shall also use the instance $I_{n,k} := \widehat{I}_{n,k}(1, X_1, \ldots, X_n)$ (which, by the way, is FdB, since $I_{n,k} = \Phi_n(\eta_k)$ with $\eta_k(x) := ((1+x)^{-(k+1)} - 1)/(k+1)$).

Theorem 7.8. *Let φ, ψ be invertible functions. Then, for every $n \geq 0$*

$$\Lambda_n(\iota, \varphi \mid \iota, \psi) = \sum_{k=0}^{n} \left(A_{n,k}^{\psi}(0) \sum_{j=0}^{k} B_{k,j}^{\varphi}(0) \, \widehat{I}_{j,k} \right).$$

Proof. Suppose the situation (7.4) with $a = b = \iota$, which implies $a_n = b_n = \delta_{n1}$, $c \in \mathcal{F}_1$ and $f_0 = 0$. Similar to the proof of Theorem 7.7, we get by Proposition 7.4

$$\overline{\Gamma}_n^{(2)}(\iota, \psi) = \sum_{k=0}^{n} A_{n,k}^{\psi}(0) \cdot \frac{1}{k+1} X_{k+1}, \tag{7.8}$$

hence by Theorem 7.6

$$\Lambda_n(\iota, \varphi \mid \iota, \psi) = \sum_{k=0}^{n} A_{n,k}^{\psi}(0) \cdot \frac{1}{k+1} A_{k+1,1}(\Gamma_1(\iota, \varphi), \ldots, \Gamma_{k+1}(\iota, \varphi)). \tag{7.9}$$

We now set $g = c \circ \varphi$ and thus obtain by Proposition 7.2 for every $n \geq 1$

$$f_n = \Gamma_n(\iota, \varphi)(c_0, \ldots, c_n) = \sum_{k=0}^{n} \left(\sum_{j=k}^{n} \binom{n}{j} \delta_{(n-j)1} B_{j,k}^{\varphi}(0) \right) c_k$$

$$= n \sum_{k=0}^{n-1} B_{n-1,k}^{\varphi}(0) c_k$$

$$= n \Omega_{n-1}(g \mid c)^c(0)$$

$$= n g_{n-1}.$$

Combining this result with (7.9) gives

$$\Lambda_n(\iota, \varphi \mid \iota, \psi)(c_0, \ldots, c_n) = \sum_{k=0}^{n} A_{n,k}^{\psi}(0) \frac{A_{k+1,1}(g_0, 2g_1, \ldots, (k+1)g_k)}{k+1}.$$

$$\tag{7.10}$$

Note that $g_0 = c_0$; then by Proposition 5.7

$$A_{k+1,1}(g_0, 2g_1, \ldots, (k+1)g_k) = \widehat{P}_{k,-(k+1)}(g_0, g_1, \ldots, g_k)$$

$$= \sum_{j=0}^{k} (-k-1)_j \, c_0^{-k-1-j} B_{k,j}^g(0).$$

Here we can evaluate $B_{k,j}^g(0)$ with the help of Jabotinsky's formula (Theorem 4.7 (i)) as follows:

$$B_{k,j}^g(0) = B_{k,j}^{c \circ \varphi}(0) = \sum_{i=j}^{k} B_{k,i}^\varphi(0) B_{i,j}(c_1, \ldots, c_{i-j+1}).$$

With this the fractional expression in (7.10) becomes

$$\frac{A_{k+1,1}(g_0, 2g_1, \ldots)}{k+1} = \sum_{j=0}^{k} \sum_{i=j}^{k} \frac{(-k-1)_j}{k+1} \, c_0^{-k-1-j} B_{k,i}^\varphi(0) B_{i,j}(c_1, c_2, \ldots).$$

Rearranging the double series and observing

$$\frac{(-k-1)_j}{k+1} = (-1)^j (k+j)_{j-1}$$

we obtain

$$\frac{A_{k+1,1}(g_0, 2g_1, \ldots, (k+1)g_k)}{k+1} = \sum_{i=0}^{k} B_{k,i}^\varphi(0) \widehat{I}_{i,k}(c_0, c_1, \ldots, c_k).$$

Substitution into (7.10) finally gives the desired result. $\diamond$

We conclude this section by showing that Theorem 7.1 (= Comtet's Theorem F) can be obtained as a corollary from Theorem 7.8.

Proof. Considering the assumptions of Theorem 7.1, one might be tempted at first glance to directly evaluate $\Lambda_n(\iota, \iota^s \,|\, \iota, \iota^s)$. This of course fails, because ι^s ($s \geq 2$) is not invertible. However, Theorem 7.8 remains valid even with the condition $\varphi \in \mathcal{G}$ weakened to $\varphi \in \mathcal{F}_0$. Therefore, as an alternative, we will first deal with calculating $\Lambda_n(\iota, \varphi \,|\, \iota, \psi)$, where $\varphi = \iota^s (\in \mathcal{F}_0)$ and $\psi = \iota (\in \mathcal{G})$. Theorem 7.8 yields

$$\Lambda_n(\iota, \iota^s \,|\, \iota, \iota) = \sum_{k=0}^{n} \left(\delta_{nk} \sum_{j=0}^{k} B_{k,j}^{\iota^s}(0) \, \widehat{I}_{j,k} \right) = \sum_{j=0}^{n} B_{n,j}^{\iota^s}(0) \, \widehat{I}_{j,n}. \tag{7.11}$$

Since $D^r(\iota^s)(0) = \delta_{sr}s!$, we have

$$
\begin{aligned}
B^{\iota^s}_{n,j}(0) &= B_{n,j}(0,\ldots,0,s!,0,\ldots,0) && (s!\text{ at position } s)\\
&= (s!)^j \cdot B_{n,j}(0,\ldots,0,1,0,\ldots,0) && (\text{by homogeneity})\\
&= (s!)^j \cdot \frac{(sj)!}{j!(s!)^j} && (\text{if } n = sj, \text{ else } 0; \text{ by } (2.13))\\
&= \delta_{n(sj)}\frac{(sj)!}{j!}.
\end{aligned}
$$

Thus (7.11) becomes

$$
\Lambda_n(\iota,\iota^s\,|\,\iota,\iota) = \sum_{j=0}^{n}\delta_{n(sj)}\frac{(sj)!}{j!}\,\widehat{I}_{j,n}. \tag{7.12}
$$

According to Theorem 7.6 the constants $e_n = \Lambda_n(\iota,\iota^s\,|\,\iota,\iota)(c_0,c_1,\ldots,c_n)$ are coefficients satisfying $\overline{f}(x) = \sum_{n\geq 0}\frac{e_n}{n!}x^{n+1}$. From Eq. (7.12) we immediately see that e_n is unequal to zero, if and only if n is an integral multiple of s. Assuming $n = sr$ for some integer $r \geq 0$ and observing $c_0 = 1$, we get

$$
e_{sr} = \frac{(sr)!}{r!}\widehat{I}_{r,sr}(c_0,c_1,\ldots,c_r) = \frac{(sr)!}{r!}I_{r,sr}(c_1,\ldots,c_r),
$$

hence

$$
\overline{f}(x) = \sum_{r\geq 0}\frac{e_{sr}}{(sr)!}x^{sr+1} = \sum_{r\geq 0}\frac{d_r}{r!}x^{sr+1}
$$

with

$$
\begin{aligned}
d_r &= I_{r,sr}(c_1,\ldots,c_r)\\
&= \sum_{j=0}^{r}(-1)^j\binom{sr+j}{j-1}(j-1)!B_{r,j}(c_1,\ldots,c_{r-j+1}). \qquad \Diamond
\end{aligned}
$$

Remark 7.3. The polynomials $\widehat{I}_{n,sn}$ and $I_{n,sn}$ are self-inverse, that is, we have $\widehat{I}_{n,sn}\circ\widehat{I}_{\sharp,s\sharp} = X_n$ $(n \geq 0)$ and $I_{n,sn}\circ I_{\sharp,s\sharp} = X_n$ $(n \geq 1)$.

8 Reciprocity theorems

In this final section we will be occupied by establishing reciprocity laws for several previously studied classes of polynomials. Stanley [93] pointed out that reciprocity (or 'duality between two related enumeration problems') is a 'rather vague concept' that only becomes clearer through concrete examples. This also applies to the reciprocity statements we are concerned with here. We are usually dealing with two families of polynomials (or sequences of numbers) that arise in some way from certain opposing aspects of a situation, while both families are actually united by their law of reciprocity.

According to Stanley [94, p. 15/16] the relationship between the number of k-combinations of n elements *with* repetitions and the corresponding number of combinations *without* repetitions is 'the simplest instance of a combinatorial reciprocity theorem' (which we shall make use of in the following):

$$\binom{n+k-1}{k} = (-1)^k \binom{-n}{k}. \tag{8.1}$$

Remark 8.1. Eq. (8.1) is valid for integers $k \geq 0$. We will extend it also in the case $k < 0$ by setting $\binom{-n}{k} = 0$, if $k > -n$, and $\binom{-n}{k} = \binom{-n}{-n-k}$ otherwise. Here again (8.1) can be applied because of $-n - k \geq 0$.

Another particularly typical example are the Stirling numbers of the first and second kind. Knuth [50] has repeatedly emphasized their importance and provided insightful historical comments on their reciprocity. This 'beautiful and easily remembered law of duality' can be expressed in different (equivalent) ways, such as

$$s_2(n, k) = c(-k, -n) \tag{8.2}$$

$$\text{or } s_2(n, k) = (-1)^{n-k} s_1(-k, -n) \tag{8.3}$$

'implying that there really is only one "kind" of Stirling number' [p. 412, ibid.]. In [49] Knuth has started with generalizing these relationships to sequences of coefficients belonging to arbitrary pairs of inverse functions.[4]

The first aim of this section is to find a *polynomial analogue* of this reciprocity, that is, a multivariable identity that turns into (8.2) or (8.3) through unification. From this point of view the resulting numerical identities might

---

[4] See also the enlightening hint from I. Gessel that Knuth mentioned in this context.

appear, so to speak, rather as a shadow of the rich intrinsic structure that is preserved in the corresponding original polynomial equations.

It is clear that $s_2(n, k)$ is to be understood as $B_{n,k} \circ 1$. As for the signless Stirling numbers of the first kind, we know the two options $c(n, k) = Z_{n,k} \circ 1$ (cf. Section 5.3) and $c(n, k) = C_{n,k} \circ 1$ (by Proposition 5.15 (i)). The former must be discarded for compelling reasons the reader will find discussed in [84, Remark 6.3]. According to Theorem 5.13, the latter amounts to $C_{n,k} \circ 1$ $= A_{n,k}((-1)^0, \ldots, (-1)^{n-k}) = (-1)^{n-k} s_1(n, k)$, that is, the resulting situation is the same as in Eq. (8.3). In summary, we therefore recognize the following identity as the best suitable candidate for the desired polynomial version of (8.3):

$$B_{n,k} = (-1)^{n-k} A_{-k,-n}. \tag{8.4}$$

Looking at (8.4) we face a problem that needs to be solved: *extending the index domain to the integers.* As far as (8.3) is concerned, Gould [32] [50, p. 417] was possibly the first to observe that the domain of Stirling numbers can be extended to negative values of n by using the fact that $s_1(n, n - k)$ and $s_2(n, n - k)$ are polynomials in n of degree $2k$. As a consequence, these polynomials can also be defined for arbitrary complex (and *a fortiori* negative integer) values. Details of Gould's method are described in [76, Section 14.1], where also a proof of (8.3) is given with the help of the Schlömilch-Schläfli formula for the signed Stirling numbers of the first kind.

Remark 8.2. One quickly sees that Gould's trick obviously does not work when transferred to the indices of the polynomial families in question. Alternatively, however, the following identity can be used for the same purpose:

$$B_{n,n-k} = \sum_{j=0}^{k} \binom{n}{k+j} X_1^{n-k-j} \widetilde{B}_{k+j,j}. \tag{8.5}$$

Eq. (8.5) can easily be derived from Eq. [31] in [22, p. 136] (cf. also [84, Corollary 4.5]) with a few calculations and index shifts. Based on this identity, one can obtain Eq. (8.4) as a consequence of Theorem 2.3 (= Theorem 6.1 in [84]), and also vice versa, it can be shown that Theorem 2.3 follows from Eq. (8.4). The proofs are omitted here[5] as we are going to take a quite different path.

To enlarge the domain of indices, we propose an easy and more straightforward procedure, which is based on the results obtained in Section 3. First, as

---

[5] For an outline of the arguments, cf. no. 27, Notes and supplements, Chapter II.

customary, we set the value of a void sum to zero. If n is negative, we therefore have according to (3.9) $\Phi_n(f) = 0$ and, in particular, $\widehat{P}_{n,k} = B_{n,k} = 0$. Next, let us have a look at the Corollaries 3.10 and 3.8. In both identities, the partial Bell polynomials $B_{n,k}$ and their orthogonal companions $A_{n,k}$ are expressed by certain instances of $\binom{n}{k}\widehat{P}_{n-k,k}$ and of $\binom{n-1}{k-1}\widehat{P}_{n-k,n}$, respectively. We now have nothing else to do but *redefine $B_{n,k}$ and $A_{n,k}$ by the right-hand sides of these identities under the assumption $n, k \in \mathbb{Z}$.* That's all!

Given any integer n, it is easy to check that $B_{n,k} = 0$ if and only if $k > n$ or n, k have unequal signs; the same holds for $A_{n,k}$. The original Stirling polynomials are restricted to indices $n = k = 0$ and $1 \leq k \leq n$, that is, they are equal to zero for all other values of n, k. After their redefinition the extended domain additionally includes $k \leq n \leq -1$. For example, instead of $B_{-3,-5} = 0$ (old value), we now get the new value

$$B_{-3,-5} = \binom{-3}{-3+5}\widehat{P}_{2,-5}\left(\tfrac{X_1}{1}, \tfrac{X_2}{2}, \tfrac{X_3}{3}\right) = 45X_1^{-7}X_2^2 - 10X_1^{-6}10X_3,$$

which is not new at all, since it is equal to $(-1)^{-3+5}A_{5,3}$ — evidently an instance of Eq. (8.4). The following matrix $(B_{n,k})$ with $-4 \leq n, k \leq 4$ shows, in a neighborhood of $(n, k) = (0, 0)$, the family of Stirling polynomials united by their fundamental reciprocity law:

$$
\begin{pmatrix}
\frac{1}{X_1^4} & 0 & 0 & 0 & 0 & 0 & 0 & 0 & 0 \\
\frac{6X_2}{X_1^5} & \frac{1}{X_1^3} & 0 & 0 & 0 & 0 & 0 & 0 & 0 \\
\frac{15X_2^2}{X_1^6} - \frac{4X_3}{X_1^5} & \frac{3X_2}{X_1^4} & \frac{1}{X_1^2} & 0 & 0 & 0 & 0 & 0 & 0 \\
\frac{15X_2^3}{X_1^7} - \frac{10X_3X_2}{X_1^6} + \frac{X_4}{X_1^5} & \frac{3X_2^2}{X_1^5} - \frac{X_3}{X_1^4} & \frac{X_2}{X_1^3} & \frac{1}{X_1} & 0 & 0 & 0 & 0 & 0 \\
0 & 0 & 0 & 0 & 1 & 0 & 0 & 0 & 0 \\
0 & 0 & 0 & 0 & 0 & X_1 & 0 & 0 & 0 \\
0 & 0 & 0 & 0 & 0 & X_2 & X_1^2 & 0 & 0 \\
0 & 0 & 0 & 0 & 0 & X_3 & 3X_1X_2 & X_1^3 & 0 \\
0 & 0 & 0 & 0 & 0 & X_4 & 3X_2^2 + 4X_1X_3 & 6X_1^2X_2 & X_1^4
\end{pmatrix}
$$

Let us now turn to the proof of the reciprocity law (8.4).

Proof. If n, k have different signs, then (8.4) holds because of $B_{n,k} = 0$ and $A_{-k,-n} = 0$. For $n = k = 0$ both sides of (8.4) are equal to 1. Suppose now that n, k are either both positive or both negative integers. In the case $k > n$ we then have $\widehat{P}_{n-k,\pm k}(\cdots) = 0$ because of $n - k < 0$ and therefore $B_{n,k} = 0 = A_{-k,-n}$ (according to the redefinition by the right-hand side expressions in Corollary 3.8 and Corollary 3.10, respectively). Thus the two cases $1 \leq k \leq n$ and $k \leq n \leq -1$ remain. Assuming the latter we obtain $n - k \geq 0$, and hence

$$B_{n,k} = \binom{n}{n-k} \widehat{P}_{n-k,k}(\tfrac{X_1}{1}, \tfrac{X_2}{2}, \ldots) \qquad \text{(Corollary 3.10, Remark 8.1)}$$

$$= (-1)^{n-k} \binom{-k-1}{n-k} \widehat{P}_{n-k,k}(\tfrac{X_1}{1}, \tfrac{X_2}{2}, \ldots) \qquad \text{(Eq. (8.1))}$$

$$= (-1)^{n-k} \binom{-k-1}{-n-1} \widehat{P}_{n-k,-k}(\widehat{R}_0(\tfrac{X_1}{1}), \widehat{R}_1(\tfrac{X_1}{1}, \tfrac{X_2}{2}), \ldots) \quad \text{(Cor. 4.2 (ii))}$$

$$= (-1)^{n-k} A_{-k,-n}. \qquad \text{(Corollary 3.8)}$$

The case $1 \leq k \leq n$ can be done in practically the same way and may be left to the reader. $\diamond$

Of course, Eq. (8.3) immediately follows from Eq. (8.4) through unification. On the other hand, it is easy to generalize Eq. (8.4) to a reciprocity theorem for B-representable polynomials.

Theorem 8.1 (General Reciprocity Law). *Let $(Q_{n,k})$ be any regular B-representable family of polynomials. Then for all $n, k \in \mathbb{Z}$*

$$Q_{n,k}^{\perp} = (-1)^{n-k} Q_{-k,-n}.$$

Proof. The above redefinitions of $B_{n,k}$ and $A_{n,k}$ can obviously be regarded as applying also to $Q_{n,k} = B_{n,k} \circ Q_{\sharp,1}$ and $Q_{n,k}^{\perp} = A_{n,k} \circ Q_{\sharp,1}$, respectively. Thus Eq. (8.4) immediately yields

$$Q_{n,k}^{\perp} = A_{n,k} \circ Q_{\sharp,1} = (-1)^{n-k} B_{-k,-n} \circ Q_{\sharp,1} = (-1)^{n-k} Q_{-k,-n}. \quad \diamond$$

Examples 8.1. In Section 5 we examined a handful of regular polynomial families, all of which obey this law of reciprocity: $Z_{n,k}$ (cycle indicators), $W_{n,k}$ (forest polynomials), $L_{n,k}$ (signed Lah polynomials, and unsigned: $L_{n,k}^{+}$) as well as $C_{n,k}$ (Comtet's polynomials). Because of $L_{n,k}^{\perp} = L_{n,k}$ there is also a case of self-reciprocity: $L_{n,k} = (-1)^{n-k} L_{-k,-n}$.

Looking back for a moment at the proof of Eq. (8.4) and at the statements of Corollaries 3.8 and 3.10, it quickly becomes clear that $\widehat{P}_{n,k}$ is the real hero of the story. In the remainder of this section, we shall deepen that impression and deal with some other interesting reciprocity properties of the potential polynomials.

Proposition 8.2. *Let* n, k *be integers with* $k \leq n \neq 0$. *Then*

$$\widehat{P}_{n-k,k}\left(\tfrac{X_1}{1}, \ldots, \tfrac{X_{n-k+1}}{n-k+1}\right) = \frac{k}{n} \cdot \widehat{P}_{n-k,-n}\left(\tfrac{A_{1,1}}{1}, \ldots, \tfrac{A_{n-k+1,1}}{n-k+1}\right).$$

Proof. Using the redefinition of the Stirling polynomials via Corollaries 3.10 and 3.8 we have

$$\widehat{P}_{n-k,k}\left(\tfrac{X_1}{1}, \ldots, \tfrac{X_{n-k+1}}{n-k+1}\right) = \binom{n}{k}^{-1} \cdot B_{n,k}$$

$$= \frac{k!}{n!}(n-k)! A_{n,k}(A_{1,1}, \ldots, A_{n-k+1,1}) \qquad \text{(by (2.15))}$$

$$= \frac{k}{n}\binom{n-1}{k-1}^{-1} \cdot A_{n,k}(A_{1,1}, \ldots, A_{n-k+1,1})$$

$$= \frac{k}{n} \cdot \widehat{P}_{n-k,n}\left(\widehat{R}_0(\tfrac{A_{1,1}}{1}), \ldots, \widehat{R}_{n-1}(\tfrac{A_{1,1}}{1}, \ldots, \tfrac{A_{n,1}}{n})\right)$$

$$= \frac{k}{n} \cdot \widehat{P}_{n-k,-n}\left(\tfrac{A_{1,1}}{1}, \ldots, \tfrac{A_{n-k+1,1}}{n-k+1}\right). \qquad \text{(Cor. 4.2 (ii))} \quad \Diamond$$

Remark 8.3. Proposition 8.2 may be regarded as a reformulation of a reciprocity law known as the *Schur-Jabotinsky theorem* (see Jabotinsky [41] and Gessel [30]). To see this we have to temporarily allow (formal) Laurent series over $\mathcal{K}$ as functions. Suppose $\varphi(x) = \sum_{n \geq 1} \varphi_n \tfrac{x^n}{n!} \in \mathcal{G}$ and $k \in \mathbb{Z}$; then the kth powers of φ and $\overline{\varphi}$ can be expanded into such series: $\varphi(x)^k = \sum_n a_{n,k} x^n$ and $\overline{\varphi}(x)^k = \sum_n b_{n,k} x^n$. Since $a_{n,k} = b_{n,k} = 0$ for $n < k$, we assume $n \geq k$. In the case $k \geq 0$ the Taylor coefficient $n! a_{n,k} = n![x^n]\varphi(x)^k$ is equal to $k! B_{n,k}^{\varphi}(0)$, whence by Corollary 3.10

$$a_{n,k} = \frac{1}{(n-k)!} \widehat{P}_{n-k,k}\left(\tfrac{\varphi_1}{1}, \ldots, \tfrac{\varphi_{n-k+1}}{n-k+1}\right).$$

This formula represents $a_{n,k}$ also for $n, k \in \mathbb{Z}$ with $n \geq k$ and can thus be applied to $\overline{\varphi}^k$:

$$b_{-k,-n} = \frac{1}{(-k+n)!} \widehat{P}_{-k+n,-n}\left(\tfrac{\overline{\varphi}_1}{1}, \ldots, \tfrac{\overline{\varphi}_{n-k+1}}{n-k+1}\right).$$

Proposition 8.2 now yields the reciprocity law in question in the form given by Gessel [Eq. (2.1.11), ibid.]:

$$a_{n,k} = \frac{k}{n} b_{-k,-n} \quad (n \neq 0).$$

In the proofs of Eq. (8.4) and of Proposition 8.2, the statement (ii) of Corollary 4.2 has been used. It seems to be the simplest reciprocity law for the potential polynomials and obviously it is also true for the variant without hat:

$$P_{n,-k} = P_{n,k}(R_1, \ldots, R_n). \tag{8.6}$$

Comtet [22] has established a formula that expresses $P_{n,-k}$ as a linear combination of the $P_{n,1}, \ldots, P_{n,n}$ (see Theorem C, p. 142, ibid.). It is stated in the following proposition and given a new and straightforward proof.

Proposition 8.3. *Let n, k be integers with $n \geq 0$ and $(-k) \notin \{0, 1, \ldots, n\}$. Then*

$$P_{n,-k} = k\binom{n+k}{n} \sum_{j=0}^{n} (-1)^j \frac{1}{k+j} \binom{n}{j} P_{n,j}.$$

Proof. By Eq. (3.10) we have

$$P_{n,-k} = \sum_{j=0}^{n} (-k)_j B_{n,j} = \sum_{j=0}^{n} j! \binom{-k}{j} B_{n,j}.$$

Replacing the binomial term by its 'reciprocal' in the sense of Eq. (8.1) and applying Bertrand's formula (3.15) to $B_{n,j}$ thus yields

$$P_{n,-k} = \sum_{j=0}^{n} j!(-1)^j \binom{k-1+j}{j} \cdot \frac{1}{j!} \sum_{r=0}^{n} (-1)^{j-r} \binom{j}{r} P_{n,r}$$

$$= \sum_{r=0}^{n} \sum_{j=0}^{n} \binom{j}{r} \binom{k-1+j}{j} (-1)^r P_{n,r}. \tag{*}$$

Clearly we have $\binom{k-1+j}{j} = \binom{k-1+j}{k-1} = \frac{k}{k+j}\binom{k+j}{k}$ and by an easy inductive argument

$$\sum_{j=0}^{n} \frac{1}{k+j} \binom{k+j}{k} \binom{j}{r} = \frac{1}{k+r} \binom{k+n}{k} \binom{n}{r}.$$

Substituting this into (*) gives the assertion. $\Diamond$

Comtet proved a slightly stronger version of Proposition 8.3, where k can be a complex number. The following theorem shows how the statement may be extended in another direction.

Theorem 8.4. *Let* $m, n, k \in \mathbb{Z}$ *with* $m \geq n \geq 0$ *and* $(-k) \notin \{0, 1, \ldots, m\}$. *Then*

$$P_{n,-k} = k \binom{m+k}{m} \sum_{j=0}^{m} (-1)^j \frac{1}{k+j} \binom{m}{j} P_{n,j}.$$

Proof. We assume n to be a fixed non-negative integer and proceed by induction on m. The basis step $m = n$ is already done by Proposition 8.3. Let S_m denote the right-hand side of the induction hypothesis. For the inductive step it is then enough to show that the difference $\Delta := S_{m+1} - S_m$ is equal to zero. Applying some elementary properties of the binomial numbers yields

$$\Delta = \binom{m+k}{k-1} \sum_{j=0}^{m+1} (-1)^j \binom{m}{j} \left(\frac{k}{k+j} P_{n,j} - \frac{k+m+1}{k+j+1} P_{n,j+1} \right). \quad (*)$$

We use the abbreviations

$$a_j := (-1)^j \binom{m}{j} \frac{k}{k+j} \quad \text{and} \quad b_j := (-1)^j \binom{m}{j} \frac{k+m+1}{k+j+1}$$

and rewrite (*) as

$$\Delta = \binom{m+k}{k-1} \left(a_0 P_{n,0} + \sum_{j=0}^{m-1} (a_{j+1} - b_j) P_{n,j+1} - b_m P_{n,m+1} \right),$$

where $a_0 P_{n,0} = \delta_{n0}$ and $-b_m P_{n,m+1} = (-1)^{m+1} P_{n,m+1}$. A little calculation gives

$$a_{j+1} - b_j = (-1)^{j+1} \binom{m+1}{j+1}.$$

In summary it results

$$\binom{m+k}{k-1}^{-1} \Delta = \delta_{n0} + \sum_{j=0}^{m-1} (-1)^{j+1} \binom{m+1}{j+1} P_{n,j+1} + (-1)^{m+1} P_{n,m+1}$$

$$= \delta_{n0} + \sum_{j=1}^{m+1} (-1)^j \binom{m+1}{j} P_{n,j}$$

$$= \sum_{j=0}^{m+1} (-1)^j \binom{m+1}{j} P_{n,j}.$$

By means of the Bertrand formula (3.15) one readily verifies that the last sum is equal to $(-1)^{m+1}(m+1)!B_{n,m+1}$, which in fact vanishes for $m \geq n$. $\Diamond$

Comtet's formula for $P_{n,-k}$ in Proposition 8.3 has a striking resemblance to a well-known binomial transformation attributed to Melzak [62, 63]. Let $p(x)$ be any polynomial of degree $\leq n$. If we now simply write $p(x+k)$ instead of $P_{n,-k}$ and $p(x-j)$ instead of $P_{n,j}$, the result is Melzak's formula

$$p(x+k) = k\binom{n+k}{n}\sum_{j=0}^{n}(-1)^j\frac{1}{k+j}\binom{n}{j}p(x-j). \qquad (8.7)$$

Recently, several authors have presented new studies on this remarkable identity. Quaintance and Gould [76] devoted Chapter 7 of their monograph to the subject. Boyadzhiev [15] and Abel [2] provided extensions (concerning the degree of $p(x)$) and new proofs. The former author even traced Eq. (8.7) back up to Nielsen's treatise [75] on Bernoulli numbers.

We finally will derive an extended version of (8.7) from Theorem 8.4. This demonstrates that the successful replacement of $P_{n,-k}$ by $p(x+k)$ in Comtet's formula is not an accident. Rather, it turns out in this way that Theorem 8.4 is the more comprehensive statement.

Theorem 8.5. *Let m, n be integers with $m \geq n \geq 0$ and $f_n(x) \in \mathbb{C}[x]$ any polynomial of degree n. Then for all $k \in \mathbb{Z} \setminus \{-m, \dots, -1, 0\}$*

$$f_n(x+k) = k\binom{m+k}{m}\sum_{j=0}^{m}(-1)^j\frac{1}{k+j}\binom{m}{j}f_n(x-j).$$

Proof. We apply $\circ R_\sharp$ on both sides of the equation of Theorem 8.4. According to the basic reciprocity law (8.6) this gives

$$P_{r,k} = k\binom{m+k}{m}\sum_{j=0}^{m}(-1)^j\frac{1}{k+j}\binom{m}{j}P_{r,-j} \qquad (8.8)$$

for every non-negative $r \leq n$. Unification then yields

$$k^r = k\binom{m+k}{m}\sum_{j=0}^{m}(-1)^j\frac{1}{k+j}\binom{m}{j}(-j)^r. \qquad (8.9)$$

By assumption we have $f_n(x) = a_0 + a_1 x + \cdots + a_n x^n$, $a_n \neq 0$, and hence from (8.9)

$$f_n(k) = k \binom{m+k}{m} \sum_{j=0}^{m} (-1)^j \frac{1}{k+j} \binom{m}{j} f_n(-j). \qquad (8.10)$$

Setting $g_{n,k}(x) := f_n(x+k) - f_n(k)$ we obtain $g_{n,k}(0) = 0$ and

$$g_{n,k}(x) = \sum_{r=1}^{n} a_r((x+k)^r - k^r) = \sum_{j=1}^{n} \underbrace{\left(\sum_{r=j}^{n} \binom{r}{j} a_r k^{r-j} \right)}_{(*)} x^j. \qquad (8.11)$$

Abbreviate the inner sum $(*)$ to $b_j(k)$.

We will now show that $g_{n,k}$ satisfies the Melzak formula. We prove that there are constants $c_1, \ldots, c_n \in \mathbb{C}$ such that for all $n \geq 1$

$$P_{n,k}(c_1 x, \ldots, c_n x) = g_{n,k}(x). \qquad (8.12)$$

According to (3.14) and because of the homogeneity of the partial Bell polynomials, the left-hand side of (8.12) can be written as

$$P_{n,k}(c_1 x, \ldots, c_n x) = \sum_{j=1}^{n} j! \binom{k}{j} B_{n,j}(c_1, \ldots, c_{n-j+1}) x^j. \qquad (8.13)$$

Equating the coefficients of x^j in (8.11) and (8.13) then yields the following equation system for the constants $c_1, \ldots, c_n$:

$$B_{n,j}(c_1, \ldots, c_{n-j+1}) = \frac{b_j(k)}{j! \binom{k}{j}} \qquad (1 \leq j \leq n). \qquad (8.14)$$

Recall that according to Remark 5.1 the system (8.14) is solvable if and only if $c_1 \in \mathbb{C}$ exists such that

$$c_1^n = \frac{b_n(k)}{n!} \binom{k}{n}^{-1}. \qquad (8.15)$$

Therefore, we choose c_1 to be any of the nth roots of the right-hand side of (8.15). Then the remaining constants $c_2, \ldots, c_n$ are uniquely determined (each of them depending on k, for example, $c_n = b_1(k)/k$). We now put

$r = n$ and replace in (8.8) each X_j by $c_j x$, $1 \leq j \leq n$. Since Eq. (8.12) is true for the constants $c_1, \ldots, c_n$, we are led to

$$g_{n,k}(x) = k \binom{m+k}{m} \sum_{j=0}^{m} (-1)^j \frac{1}{k+j} \binom{m}{j} g_{n,-j}(x). \qquad (8.16)$$

Because of $f_n(x+k) = f_n(k) + g_{n,k}(x)$ the assertion follows from Eq. (8.10) and Eq. (8.16). $\diamond$

NOTES AND SUPPLEMENTS

1. Translation rule for coefficients [p. 60]. The following simple identity proves useful, when dealing with coefficients: Let f be any function from $\mathcal{F}$; then for every $n, r \in \mathbb{Z}$

$$[x^n] x^r f(x) = [x^{n-r}] f(x).$$

(Exercise)

2. Invertible product [p. 61]. We prove the statement from Remark 2.1. Suppose a, b, c are functions with coefficients $a_n = [x^n] a(x)$, $b_n = [x^n] b(x)$, and $c_n = [x^n] c(x)$, furthermore $c = a \cdot b$. Then we have $c(0) = a_0 b_0$ and $c'(0) = a_1 b_0 + a_0 b_1$.

(i) Assume c to be invertible. It holds $a_0 b_0 = 0$ and $a_1 b_0 + a_0 b_1 \neq 0$. Since a_0 and b_0 cannot both be zero, assume (without loss of generality) $a_0 \neq 0$ and $b_0 = 0$. Thus a is a unit, and it follows $a_0 b_1 \neq 0$, whence $b_1 \neq 0$, that is, b is invertible.

(ii) Suppose now conversely, say $a \in \mathcal{G}$ and $b \in \mathcal{F}_1$. Then $a_0 = 0$, $a_1 \neq 0$, and $b_0 \neq 0$. This implies $c(0) = 0$, $c'(0) \neq 0$, and hence $c \in \mathcal{G}$.

3. Exponential function [p. 61]. a) The exponential exp can be characterized as the unique solution of the differential equation $D(f) = f$ ($f \in \mathcal{F}$) under the initial value condition $f(0) = 1$.

Proof. Writing $f(x) = \sum_{n \geq 0} f_n \frac{x^n}{n!}$ we immediately obtain from $D(f)(x) = f(x)$, by equating the coefficients, $f_{n+1} = f_n$ for $n = 0, 1, 2, \ldots$

Since $f_0 = 1$, all Taylor coefficients f_n must be equal to 1.

b) For all $f, g \in \mathcal{F}_0$ the functional equation $e^f \cdot e^g = e^{f+g}$ holds[6]. (Exercise)

c) Choosing $f = -g$ in b), we get $1 = \exp(0) = \exp(-g)\exp(g)$, and hence $\exp(g)^{-1} = \exp(-g)$. More generally, $\exp(mg) = \exp(g)^m$ holds for all $m \in \mathbb{Z}$ and $g \in \mathcal{F}_0$. (Exercise)

4. Exponential and logarithm modified [p. 61]. The formal power series $\varepsilon, \lambda \in \mathcal{F}$ are algebraic versions of $e^x - 1$ and $\log(1+x)$, respectively. They are inverse functions of each other: $\varepsilon \circ \lambda = \lambda \circ \varepsilon = \iota$. This is immediately clear, since the Taylor coefficients of $\bar{\varepsilon}$ are given by $A_{n,1}^{\varepsilon}(0) = A_{n,1}(1, \ldots, 1) = s_1(n, 1) = (-1)^{n-1}(n-1)!$, and hence

$$\bar{\varepsilon}(x) = \sum_{n \geq 1} (-1)^{n-1}(n-1)! \frac{x^n}{n!} = \lambda(x).$$

5. Computing inverse functions [p. 63]. Let φ be any function from $\mathcal{G}$; then the Taylor coefficients of the inverse function $\overline{\varphi}$ can be represented by the Lagrangian term $D^{n-1}((\iota/\varphi)^n)(0)$ as well as by the nth derivative of ι with respect to φ (at zero), that is, by $D_\varphi^n(\iota)(0)$.

Which of the two expressions is computationally more efficient? It would be worthwhile looking for a criterion that provides answers to this question depending on certain classes of functions. The problem becomes even more interesting when the assumption will be dropped that functions are given in the form of a power series.

A prominent example of this kind is $\varphi(x) = xe^{-x}$ (cf. Chapter I, Example 7.1). Obviously, this function seems to be tailored to the application of the Lagrange method, for we immediately have $(\iota/\varphi)^n(x) = e^{nx}$, and hence $D^{n-1}((\iota/\varphi)^n)(0) = n^{n-1}$. Compared with this, the calculation of $D_\varphi^n(\iota)$ requires significantly more effort[7]. However, examples of this type appear to be more of an exception. In most cases, calculating ›inverse coefficients‹ by means of the Lagrange formula turns out to be rather complicated. This is already indicated if one takes into account the respective general expressions for small values of the order n.

---

[6] where $e^h := \exp(h) := \exp \circ h$

[7] On obtains $D_\varphi^n(\iota)(x) = e^{nx}(x-1)^{-(2n-1)}p_n(x)$, where $p_n(x)$ is a polynomial of degree n with $p_n(0) = -n^{n-1}$.

For example, taking $n = 4$ we get

$$D^3((\iota/\varphi)^4)(x) = \frac{24x}{\varphi(x)^4} - \frac{144x^2\varphi'(x)}{\varphi(x)^5} + \frac{240x^3\varphi'(x)^2}{\varphi(x)^6} - \frac{120x^4\varphi'(x)^3}{\varphi(x)^7}$$
$$- \frac{48x^3\varphi''(x)}{\varphi(x)^5} + \frac{60x^4\varphi'(x)\varphi''(x)}{\varphi(x)^6} - \frac{4x^4\varphi^{(3)}(x)}{\varphi(x)^5}.$$

Note that because of $\varphi(0) = 0$ the right-hand side may possibly not be evaluated directly for $x = 0$; one then would have to carry out $x \longrightarrow 0$ with the help of L'Hôpital's rule.

On the other hand, the iterative term

$$D_\varphi^4(\iota)(x) = -\frac{15\varphi''(x)^3}{\varphi'(x)^7} + \frac{10\varphi''(x)\varphi^{(3)}(x)}{\varphi'(x)^6} - \frac{\varphi^{(4)}(x)}{\varphi'(x)^5}$$

avoids this detour and also has less than half of the summands. The example

$$\varphi(x) = \arctan(x) = \int_0^x \frac{dt}{1+t^2}$$

may be used here to illustrate how these conditions affect practical computing. While the Lagrange method delivers unresolved expressions (as above), the terms $d_n(x) := D_{\arctan}^n(\iota)(x)$ $(n = 1, 2, 3, \ldots)$ can be calculated using the iterative rule

$$d_1(x) = 1 + x^2$$
$$d_{n+1}(x) = (1 + x^2)d_n'(x).$$

One easily obtains

$$d_2(x) = 2x + 2x^3$$
$$d_3(x) = 2 + 8x^2 + 6x^4$$
$$d_4(x) = 16x + 40x^3 + 24x^5$$
$$\text{etc.}$$

and thus the Taylor coefficients of $\tan(x)$: $1, 0, 2, 0, 16, 0, 272, 0, 7936, 0, \ldots$

It can be left to the reader to find out the connection between the Bernoulli numbers appearing in the tan-series and the (exponential) Lagrange inversion polynomial $A_{n,1}$ (cf. no. 17, Notes and supplements, Chapter I).

6. Todorov's determinant [p. 64].   Todorov [99, Theorem 6] expresses the higher derivatives of f with respect to φ by a determinant. Suppose $1 \le j \le n$ and $1 \le k \le \min(j, n-1)$; furthermore set

$$t_{j,k} := \frac{((j-k+1)n - j) \cdot D^{j-k+1}(\varphi)}{(j-k+1)!}.$$

Then the following holds for every $n \ge 1$:

$$D_{\varphi}^n(f) = \frac{1}{D(\varphi)^{2n-1}} \begin{vmatrix} t_{1,1} & 0 & \cdots & 0 & D^1(f)/0! \\ t_{2,1} & t_{2,2} & \cdots & 0 & D^2(f)/1! \\ \vdots & \vdots & \vdots & \vdots & \vdots \\ t_{n-1,1} & t_{n-1,2} & \cdots & t_{n-1,n-2} & D^{n-1}(f)/(n-2)! \\ t_{n,1} & t_{n,2} & \cdots & t_{n,n-1} & D^n(f)/(n-1)! \end{vmatrix}$$

Expanding the determinant with respect to the nth column we would obtain a formula similar to that in Proposition 3.1 (Chapter I), with the difference that the coefficients $a_{n,k}$ of $D^k(f)$ are not defined by recursion, but again by determinants.

7. Multiplication rule [p. 65].   Suppose that $t \in \mathcal{K}$ and $a, b \in \mathbb{Z}$. Then we have

$$B_{n,k}(t^{a+b}X_1, t^{2a+b}X_2, t^{3a+b}X_3, \ldots) = t^{an+bk}B_{n,k}.$$

(Exercise) — The following special cases are frequently used statements:

$$B_{n,k}(X_1, tX_2, t^2X_3, \ldots) = t^{n-k}B_{n,k}$$
$$B_{n,k}(tX_1, t^2X_2, t^3X_3, \ldots) = t^n B_{n,k}.$$

Setting $t = -1$ one obtains useful sign-rules.

8. Some specializations [p. 66].   The following statements about the modified exponential ε and logarithm λ correspond to the identities given in Examples 3.1 (i), Chapter I, with regard to exp and log.

(1) $A_{n,k}^{\varepsilon} = s_1(n,k) \cdot (1+\varepsilon)^{-n}$ (1') $B_{n,k}^{\lambda} = s_1(n,k) \cdot (1+\iota)^{-n}$

$$(2) \qquad B_{n,k}^{\varepsilon} = s_2(n,k) \cdot (1+\varepsilon)^k \qquad (2') \qquad A_{n,k}^{\lambda} = s_2(n,k) \cdot (1+\iota)^k.$$

Proof of (2) and (2'):
$$
\begin{aligned}
B_{n,k}^{\varepsilon} &= B_{n,k}(D(\varepsilon), \ldots, D^{n-k+1}(\varepsilon)) \\
&= B_{n,k}(\exp, \ldots, \exp) \\
&= \exp^k B_{n,k}(1, \ldots, 1) \\
&= s_2(n,k)(1+\varepsilon)^k.
\end{aligned}
$$

$$
\begin{aligned}
A_{n,k}^{\lambda} &= B_{n,k}^{\overline{\lambda}} \circ \lambda \\
&= B_{n,k}^{\varepsilon} \circ \lambda \\
&= (s_2(n,k) \circ \lambda) \cdot (1 \circ \lambda + \varepsilon \circ \lambda)^k \\
&= s_2(n,k)(1+\iota)^k.
\end{aligned}
$$

(1) and (1') are left to the reader.

9. Table: Reciprocal polynomials [p. 70].

$$
\widehat{R}_0 = \frac{1}{X_0},
$$

$$
\widehat{R}_1 = -\frac{X_1}{X_0^2}
$$

$$
\widehat{R}_2 = \frac{2X_1^2}{X_0^3} - \frac{X_2}{X_0^2}
$$

$$
\widehat{R}_3 = -\frac{6X_1^3}{X_0^4} + \frac{6X_2X_1}{X_0^3} - \frac{X_3}{X_0^2}
$$

$$
\widehat{R}_4 = \frac{24X_1^4}{X_0^5} - \frac{36X_2X_1^2}{X_0^4} + \frac{8X_3X_1}{X_0^3} + \frac{6X_2^2}{X_0^3} - \frac{X_4}{X_0^2}
$$

$$
\widehat{R}_5 = -\frac{120X_1^5}{X_0^6} + \frac{240X_2X_1^3}{X_0^5} - \frac{60X_3X_1^2}{X_0^4} - \frac{90X_2^2X_1}{X_0^4} + \\
\frac{10X_4X_1}{X_0^3} + \frac{20X_2X_3}{X_0^3} - \frac{X_5}{X_0^2}
$$

10. Reciprocal polynomials [p. 70].

The polynomials $\widehat{R}_n$ can be expressed by means of R_n as follows:

$$
\widehat{R}_n = \frac{1}{X_0} R_n\left(\frac{X_1}{X_0}, \ldots, \frac{X_n}{X_0}\right). \qquad \text{(Exercise)}
$$

11. Table: Tree polynomials [p. 71].

$$\widehat{T}_1 = X_0$$
$$\widehat{T}_2 = 2X_0X_1$$
$$\widehat{T}_3 = 3X_2X_0^2 + 6X_1^2X_0$$
$$\widehat{T}_4 = 4X_3X_0^3 + 36X_1X_2X_0^2 + 24X_1^3X_0$$
$$\widehat{T}_5 = 5X_4X_0^4 + 60X_2^2X_0^3 + 80X_1X_3X_0^3 + 360X_1^2X_2X_0^2 + 120X_1^4X_0$$
$$\widehat{T}_6 = 6X_5X_0^5 + 300X_2X_3X_0^4 + 150X_1X_4X_0^4 + 1800X_1X_2^2X_0^3$$
$$+ 1200X_1^2X_3X_0^3 + 3600X_1^3X_2X_0^2 + 720X_1^5X_0$$

12. Table: Logarithmic polynomials [p. 72].

$$L_1 = X_1$$
$$L_2 = X_2 - X_1^2$$
$$L_3 = 2X_1^3 - 3X_2X_1 + X_3$$
$$L_4 = -6X_1^4 + 12X_2X_1^2 - 4X_3X_1 - 3X_2^2 + X_4$$
$$L_5 = 24X_1^5 - 60X_2X_1^3 + 20X_3X_1^2 + 30X_2^2X_1 - 5X_4X_1 - 10X_2X_3 + X_5$$
$$L_6 = -120X_1^6 + 360X_2X_1^4 - 120X_3X_1^3 - 270X_2^2X_1^2 + 30X_4X_1^2$$
$$+ 120X_2X_3X_1 - 6X_5X_1 + 30X_2^3 - 10X_3^2 - 15X_2X_4 + X_6$$

13. Table: Potential polynomials [p. 72].

The first 5 instances of the 5th generation of the potential polynomial family:

$$\widehat{P}_{5,1} = X_5$$
$$\widehat{P}_{5,2} = 20X_2X_3 + 10X_1X_4 + 2X_0X_5$$
$$\widehat{P}_{5,3} = 3X_5X_0^2 + 60X_2X_3X_0 + 30X_1X_4X_0 + 90X_1X_2^2 + 60X_1^2X_3$$
$$\widehat{P}_{5,4} = 4X_5X_0^3 + 120X_2X_3X_0^2 + 60X_1X_4X_0^2 + 360X_1X_2^2X_0$$
$$+ 240X_1^2X_3X_0 + 240X_1^3X_2$$
$$\widehat{P}_{5,5} = 120X_1^5 + 1200X_0X_2X_1^3 + 600X_0^2X_3X_1^2 + 900X_0^2X_2^2X_1$$
$$+ 100X_0^3X_4X_1 + 200X_0^3X_2X_3 + 5X_0^4X_5$$

14. Factorial polynomials and their coefficient sums [p. 73].

a) The first 4 instances of the 4th generation of the factorial polynomial family:

$$\widehat{F}_{4,1} = X_4$$

$$\widehat{F}_{4,2} = 6X_2^2 + 8X_1X_3 + 2X_0X_4 - X_4$$

$$\widehat{F}_{4,3} = 3X_4X_0^2 + 18X_2^2X_0 + 24X_1X_3X_0 - 6X_4X_0 - 18X_2^2 + 36X_1^2X_2$$
$$- 24X_1X_3 + 2X_4$$

$$\widehat{F}_{4,4} = 24X_1^4 + 144X_0X_2X_1^2 - 216X_2X_1^2 + 48X_0^2X_3X_1 - 144X_0X_3X_1$$
$$+ 88X_3X_1 + 36X_0^2X_2^2 - 108X_0X_2^2 + 66X_2^2 + 4X_0^3X_4$$
$$- 18X_0^2X_4 + 22X_0X_4 - 6X_4$$

b) The sequence of numbers $[n]_k = \widehat{F}_{n,k}(1,\dots,1)$ satisfies the following identity (cf. Remark 3.3 (v))[8]:

$$[n]_k = \sum_{j=1}^{\min(k,n)} (-1)^{k-j} j! s_2(n,j) \left(\begin{bmatrix} k-1 \\ j-1 \end{bmatrix} - \begin{bmatrix} k-1 \\ j \end{bmatrix} \right). \tag{*}$$

Proof (Outline): Recall that

$$r^n = \sum_{j=0}^{n} j! \binom{r}{j} s_2(n,j) \tag{1}$$

and $\quad [n]_k = \widehat{F}_{n,k} \circ 1 = \sum_{r=1}^{k} s_1(k,r) r^n. \tag{2}$

Substitution of (1) into (2) gives

$$[n]_k = \sum_{j=1}^{\min(k,n)} j! s_2(n,j) \sum_{r=j}^{k} \binom{r}{j} s_1(k,r). \tag{3}$$

For the next step, the following upper summation formula is required:

$$\sum_{r=j+1}^{k} \binom{r}{j} s_1(k,r) = k s_1(k-1,j). \tag{4}$$

---

[8] using Karamata's notation recommended by Knuth for the cycle numbers $c(n,k)$

(4) can be proved by induction on k. (Exercise)

It follows from (4)

$$\sum_{r=j}^{k} \binom{r}{j} s_1(k,r) = s_1(k,r) + k s_1(k-1,r)$$
$$= s_1(k-1,r-1) + s_1(k-1,r)$$
$$= (-1)^{k-r} c(k-1,r-1) - (-1)^{k-r} c(k-1,r).$$

This, together with (3), yields the asserted equation (*).   ◇

c) Table of numbers $[n]_k$ for $1 \leq n, k \leq 7$:

$k =$	1	2	3	4	5	6	7
$n = 1$	1	1	-1	2	-6	24	-120
$n = 2$	1	3	-1	0	4	-28	188
$n = 3$	1	7	5	-16	54	-222	1098
$n = 4$	1	15	35	-60	124	-280	440
$n = 5$	1	31	149	-88	-186	1914	-13350
$n = 6$	1	63	539	420	-2996	13832	-66592
$n = 7$	1	127	1805	4664	-15546	43578	-98862

d) Finally, let us consider some instances of (*) for small exponents ($n = 1, 2, 3$) thus demonstrating that the power sum indeed undergoes a substantial simplification through the right-hand expression of (*).

Let $n \geq 2$; then, cancelling $(-1)^k$ in each equation we obtain the identities

$$\sum_{r=1}^{k} (-1)^r \begin{bmatrix} k \\ r \end{bmatrix} r = \begin{bmatrix} k-1 \\ 1 \end{bmatrix},$$

$$\sum_{r=1}^{k} (-1)^r \begin{bmatrix} k \\ r \end{bmatrix} r^2 = 3 \begin{bmatrix} k-1 \\ 1 \end{bmatrix} - 2 \begin{bmatrix} k-1 \\ 2 \end{bmatrix},$$

$$\sum_{r=1}^{k} (-1)^r \begin{bmatrix} k \\ r \end{bmatrix} r^3 = 7 \begin{bmatrix} k-1 \\ 1 \end{bmatrix} - 12 \begin{bmatrix} k-1 \\ 2 \end{bmatrix} + 6 \begin{bmatrix} k-1 \\ 3 \end{bmatrix}.$$

e) Boyadzhiev [12] has proved an identity analogous to (*), in which instead of $s_1(k, j)$ the cycle numbers $c(k, j) = |s_1(k, j)|$ were used:

$$\sum_{j=1}^{k} \begin{bmatrix} k \\ j \end{bmatrix} j^n = \sum_{j=1}^{k} j! s_2(n, j) \begin{bmatrix} k+1 \\ j+1 \end{bmatrix}.$$

See also Hsu [37].

15. Dual companions [p. 82]. Let F_n denote the FdB polynomial $\Phi_n(f) = \Omega_n(f \circ \varphi \,|\, \varphi)$ associated with a function $f \in \mathcal{F}$. We then define the *dual companion* F_n^* of F_n to be $\Phi_n^*(f) := \Omega_n(f \circ \overline{\varphi} \,|\, \varphi)$. By Proposition 3.2 we have $\Omega_n(\overline{\varphi} \,|\, \varphi) = A_{n,1}$; therefore Eq. (3.6) yields

$$\Omega_n(f \circ \overline{\varphi} \,|\, \varphi) = \sum_{k=0}^{n} D^k(f)(0) \cdot (B_{n,k} \circ A_{\sharp,1}),$$

and hence

$$F_n^* = F_n \circ A_{\sharp,1} = F_n(A_{1,1}, A_{2,1}, \ldots).$$

Clearly one has $F_n^{**} = F_n$, $X_n^* = A_{n,1}$, and the substitution rule

$$F_n(H_1, H_2, \ldots)^* = F_n(H_1^*, H_2^*, \ldots).$$

As a consequence, equations (2.14) and (2.15) may be expressed by saying that $A_{n,k}$ and $B_{n,k}$ are dual companions of each other. So, the question arises as to which orthogonal pairs are at the same time pairs of dual companions. The answer is stated in the following

Theorem. *For any regular B-representable family* $(Q_{n,k})$ *of FdB polynomials holds:* $Q_{n,k}^{\perp} = Q_{n,k}^*$ *if and only if* $Q_{n,k} = B_{n,k}$.

We give at least a rough idea of how to prove the non-trivial part (›only if‹). Using Proposition 5.1 (iv), the assumption $Q_{n,k}^{\perp} = Q_{n,k}^*$ can be written as

$$A_{n,k}(Q_{1,1}, \ldots, Q_{n-k+1,1}) = Q_{n,k}(A_{1,1}, \ldots, A_{n-k+1,1}). \tag{†}$$

According to Proposition 5.2 there exist $a_1, a_2, \ldots, a_n \in \mathcal{K}$ such that

$$Q_{n,k} = \sum_{j=k}^{n} B_{j,k}(a_1, \ldots, a_{j-k+1}) B_{n,j}$$

for $1 \leq k \leq n$. Thus (†) can be regarded as a system of equations for the unknowns $a_1, a_2, \ldots, a_n$. We are to show that $a_1 = 1$ and $a_2 = \cdots = a_n = 0$, for this implies

$$Q_{n,k} = \sum_{j=k}^{n} B_{j,k}(1,0,\ldots,0)B_{n,j} = \sum_{j=k}^{n} \delta_{jk}B_{n,j} = B_{n,k}.$$

Hint: Start with discussing the cases $k = n$, $k = n - 1$ (and so forth).   ◊

Note that also $Q_{n,k} = A_{n,k}$ satisfies (†), but as a dual companion of some FdB polynomial (without being itself FdB).

16. Polynomials not B-representable [p. 84].   Let $(Q_{n,k})$ be any B-representable family of FdB polynomials. Then by Proposition 5.1 (iii) one obtains $Q_{n,n} = c^n X_1^n$, for some $c \in \mathcal{K}$. (Exercise). — Using this as a necessary condition for B-representability, it follows immediately that the polynomials $\widehat{P}_{n,k}$, $P_{n,k}$, $\widehat{F}_{n,k}$, $F_{n,k}$ are not B-representable.

17. Table: Partial cycle indicator polynomials [p. 88].   The 7th generation of the partial cycle indicator polynomials:

$$Z_{7,1} = 720X_7$$
$$Z_{7,2} = 420X_3X_4 + 504X_2X_5 + 840X_1X_6$$
$$Z_{7,3} = 504X_5X_1^2 + 280X_3^2X_1 + 630X_2X_4X_1 + 210X_2^2X_3$$
$$Z_{7,4} = 210X_4X_1^3 + 420X_2X_3X_1^2 + 105X_2^3X_1$$
$$Z_{7,5} = 70X_3X_1^4 + 105X_2^2X_1^3$$
$$Z_{7,6} = 21X_1^5X_2$$
$$Z_{7,7} = X_1^7$$

18. Table: Forest polynomials [p. 90].   The 6th generation of the forest polynomials:

$$W_{6,1} = X_5X_0^5 + 300X_2X_3X_0^4 + 150X_1X_4X_0^4 + 1800X_1X_2^2X_0^3$$
$$+ 1200X_1^2X_3X_0^3 + 3600X_1^3X_2X_0^2 + 720X_1^5X_0$$
$$W_{6,2} = 30X_4X_0^5 + 450X_2^2X_0^4 + 600X_1X_3X_0^4 + 3600X_1^2X_2X_0^3 + 1800X_1^4X_0^2$$
$$W_{6,3} = 60X_3X_0^5 + 900X_1X_2X_0^4 + 1200X_1^3X_0^3$$

$$W_{6,4} = 60X_2X_0^5 + 300X_1^2X_0^4$$
$$W_{6,5} = 30X_0^5X_1$$
$$W_{6,6} = X_0^6$$

19. Signed Lah polynomials [p. 91]. (Exercise). — Show that

$$L_{n,n} = (-1)^n$$

$$L_{n,n-1} = (-1)^n \cdot 2\binom{n}{2}\frac{X_2}{X_1^2}$$

$$L_{n,n-2} = (-1)^n \cdot 2\binom{n-1}{2}\binom{n}{2}\frac{X_2^2}{X_1^4}.$$

20. Involutory functions [p. 94]. According to Proposition 5.11 every involution $f \in \mathcal{G}$ can be written in the form

$$f(x) = -x + L_{2,1}(c_1, c_2)\frac{x^2}{2!} + L_{3,1}(c_1, c_2)\frac{x^3}{3!} + L_{4,1}(c_1, c_2, c_3, c_4)\frac{x^4}{4!} + \cdots$$

with appropriate constants $c_1, c_2, c_3, \ldots \in \mathcal{K}$. The first Lah polynomials used to build the Taylor coefficients of f are as follows:

$$L_{1,1} = -1, \quad L_{2,1} = \frac{2X_2}{X_1^2}, \quad L_{3,1} = -\frac{6X_2^2}{X_1^4},$$

$$L_{4,1} = \frac{30X_2^3}{X_1^6} - \frac{8X_3X_2}{X_1^5} + \frac{2X_4}{X_1^4},$$

$$L_{5,1} = -\frac{210X_2^4}{X_1^8} + \frac{120X_3X_2^2}{X_1^7} - \frac{30X_4X_2}{X_1^6}.$$

Unification ($c_1 = c_2 = \ldots = 1$) yields $L_{n,1}(1, \ldots, 1) = l(n, 1) = (-1)^n n!$, and hence

$$f(x) = -x + x^2 - x^3 + x^4 - \cdots = -\frac{x}{1+x}.$$

We obtain the same result when putting $c_j = a^j$ with a fixed $a \in \mathcal{K}$. (Exercise)

21. Diophantine solutions [p. 95]. Let m, s be non-negative integers, $s \geq 1$. The number of non-negative (integral) solutions of $x_1 + \cdots + x_s = m$ is $\binom{s+m-1}{m}$. (Exercise)

In order to obtain Comtet's polynomial $C_{6,2}$ by evaluating Eq. (5.11), one needs the solutions of the diophantine equation $\rho(6,1) = r_1 + \cdots + r_5 = 6 - 2 = 4$; the number of these solutions is equal to $\binom{5+4-1}{4} = 70$.

22. Table: Comtet's polynomials [p. 97].

a) The 6th generation of Comtet's polynomials:

$$C_{6,1} = X_5 X_0^5 + 15 X_2 X_3 X_0^4 + 11 X_1 X_4 X_0^4 + 34 X_1 X_2^2 X_0^3$$
$$+ 32 X_1^2 X_3 X_0^3 + 26 X_1^3 X_2 X_0^2 + X_1^5 X_0$$
$$C_{6,2} = 6 X_4 X_0^5 + 34 X_2^2 X_0^4 + 57 X_1 X_3 X_0^4 + 146 X_1^2 X_2 X_0^3 + 31 X_1^4 X_0^2$$
$$C_{6,3} = 15 X_3 X_0^5 + 120 X_1 X_2 X_0^4 + 90 X_1^3 X_0^3$$
$$C_{6,4} = 20 X_2 X_0^5 + 65 X_1^2 X_0^4$$
$$C_{6,5} = 15 X_0^5 X_1$$
$$C_{6,6} = X_0^6$$

b) The polynomials $C_{n,n-k}$, $k = 0, 1, 2, 3$:

$$C_{n,n} = X_0^n$$

$$C_{n,n-1} = \binom{n}{2} X_0^{n-1} X_1$$

$$C_{n,n-2} = \binom{n}{3} \frac{3n - 5}{4} \cdot X_0^{n-2} X_1^2 + \binom{n}{3} X_0^{n-1} X_2,$$

$$C_{n,n-3} = \binom{n}{4} \frac{(n-2)(n-3)}{2} \cdot X_0^{n-3} X_1^3 + 2 \binom{n}{4} (n-2) X_0^{n-2} X_1 X_2$$
$$+ \binom{n}{4} X_0^{n-1} X_3$$

23. Rearranging double series [p. 103].

Whenever ›rearranging double series‹ is mentioned in the present work, the following *Standard Interchange Formula* (according to [76, p. 26]) is meant:

$$\sum_{j=0}^{n} \sum_{k=0}^{j} a_{k,j} = \sum_{k=0}^{n} \sum_{j=k}^{n} a_{k,j}.$$

24. A special inverse relation [p. 107]. Let $a, b \in \mathbb{Z}$, not both equal to 0. Birmajer, Gil and Weiner [10] have proven that the polynomials

$$U_n = \sum_{k=1}^{n} \binom{an + bk}{k - 1} (k - 1)! B_{n,k},$$

$$V_n = \sum_{k=1}^{n} \frac{an + bk}{an + b} \binom{-an - b}{k - 1} (k - 1)! B_{n,k}.$$

satisfy the inverse relation

$$U_n \circ V_\sharp = V_n \circ U_\sharp = X_n \quad \text{for all } n \geq 1.$$

For $b = 1$ this gives Comtet's Theorem F [22, p. 151], which is equivalent to the statement of Theorem 7.1 (Chapter II). As is shown in Section 7, the latter also results as a corollary from Theorem 7.8 by specializing the Lagrange inversion polynomial $\Lambda_n(\iota, \varphi \,|\, \iota, \psi)$. Comtet himself (apart from a recommendation to use Lagrange inversion) did not provide any explicit (published) proof for his Theorem F. In the next note, therefore, a direct proof will be given that does not make the statement appear merely as a corollary of more general theorems.

25. Comtet's Theorem F [p. 107]. We refer to the denotations and assumptions of Theorem 7.1. Since $f(0) = 0$ and $f'(0) = c_0 \neq 0$, the function f is invertible and has the asserted form with $d_0 = 1$. Setting now $g(x) := x/f(x)$, it follows by Lagrange inversion

$$[x^n]\overline{f}(x) = \frac{1}{n}[x^{n-1}]g(x)^n, \tag{1}$$

and hence

$$d_n = n![x^{sn+1}]\overline{f}(x) = \frac{n!}{sn + 1}[x^{sn}]g(x)^{sn+1}. \tag{2}$$

In order to calculate $g(x)^{sn+1}$, we define the auxiliary function h_s by

$$h_s(x) := \frac{f(x)}{x} - 1 = \sum_{n \geq 1} \frac{c_n}{n!} x^{sn}$$

thus obtaining

$$g(x)^{sn+1} = \left(\frac{1}{1 + h_s(x)}\right)^{sn+1}$$

$$= \sum_{k \geq 0} (-1)^k \binom{sn+k}{k} h_s(x)^k$$

$$= \sum_{k \geq 0} (-1)^k (sn+k)_k \cdot \frac{1}{k!} h_s(x)^k. \tag{3}$$

Because of $h_s(0) = 0$ we have by Theorem 3.9

$$\frac{1}{k!} h_s(x)^k = \sum_{j \geq k} B_{j,k}(h_s'(0), h_s''(0), \ldots) \frac{x^j}{j!}. \tag{4}$$

Hence, combining (2), (3), (4), and observing that $B_{sn,k}^{h_s}(0) = 0$ for $n \geq 1$, $k = 0$ or $k > sn$, we obtain

$$d_n = \frac{n!}{sn+1} \sum_{k \geq 0} (-1)^k (sn+k)_k [x^{sn}] \frac{h_s(x)^k}{k!}$$

$$= \frac{n!}{sn+1} \sum_{k \geq 0} (-1)^k (sn+k)_k B_{sn,k}^{h_s}(0) \cdot \frac{1}{(sn)!}$$

$$= \frac{n!}{(sn+1)!} \sum_{k=1}^{sn} (-1)^k (sn+k)_k B_{sn,k}^{h_s}(0). \tag{5}$$

Next we evaluate $B_{sn,k}^{h_s}(0)$ by using the simple fact that $h_s = h_1 \circ \iota^s$, and by the Jabotinsky formula (Theorem 4.7, (i)):

$$B_{sn,k}^{h_s}(0) = \sum_{j=k}^{sn} B_{sn,j}^{\iota^s}(0) B_{j,k}^{h_1}(0). \tag{6}$$

Just like in the proof of Theorem 7.1, we have

$$B_{sn,j}^{\iota^s}(0) = B_{sn,j}(0, \ldots, s!, 0, \ldots, 0) \qquad \text{(with } s! \text{ at position } s\text{)}$$

$$= \delta_{(sn)(sj)} \frac{(sj)!}{j!},$$

hence (6) becomes

$$B_{sn,k}^{h_s}(0) = \sum_{j=k}^{sn} \delta_{(sn)(sj)} \frac{(sj)!}{j!} B_{j,k}(c_1, \ldots, c_{j-k+1})$$

$$= \frac{(sn)!}{n!} B_{n,k}(c_1, \ldots, c_{n-k+1}). \tag{7}$$

Finally, substituting (7) into (5) we get

$$d_n = \frac{n!}{(sn+1)!} \sum_{k=1}^{sn} (-1)^k (sn+k)_k \frac{(sn)!}{n!} B_{n,k}(c_1, \ldots, c_{n-k+1})$$

$$= \sum_{k=1}^{n} (-1)^k (sn+k)_{k-1} B_{n,k}(c_1, \ldots, c_{n-k+1}).$$

This completes the proof. $\Diamond$

26. Table: Generalized Lagrange inversion polynomials [p. 112]. Let $a \in \mathcal{G}$ and $b \in \mathcal{F}_1$. Then the first three instances of the generalized Lagrange inversion polynomials are as follows:

$$\Lambda_1(a, \varphi \,|\, b, \psi) = \frac{1}{a_1 b_0 \psi_1 X_0}$$

$$\Lambda_2(a, \varphi \,|\, b, \psi) = -\frac{2\varphi_1 X_1}{a_1^2 b_0 \psi_1^2 X_0^3} - \frac{a_2}{a_1^3 b_0 \psi_1^2 X_0^2} - \frac{\psi_2}{a_1 b_0 \psi_1^3 X_0}$$

$$\quad - \frac{2b_1}{a_1 b_0^2 \psi_1^2 X_0}$$

$$\Lambda_3(a, \varphi \,|\, b, \psi) = \frac{9 a_2 \varphi_1 X_1}{a_1^4 b_0 \psi_1^3 X_0^4} + \frac{6 \varphi_1 \psi_2 X_1}{a_1^2 b_0 \psi_1^4 X_0^3} + \frac{12 \varphi_1^2 X_1^2}{a_1^3 b_0 \psi_1^3 X_0^5}$$

$$\quad - \frac{3 \varphi_1^2 X_2}{a_1^3 b_0 \psi_1^3 X_0^4} + \frac{6 b_1 \varphi_1 X_1}{a_1^2 b_0^2 \psi_1^3 X_0^3} - \frac{3 \psi_2 X_1}{a_1^3 b_0 \psi_1^3 X_0^4}$$

$$\quad + \frac{3 a_2^2}{a_1^5 b_0 \psi_1^3 X_0^3} + \frac{3 a_2 \psi_2}{a_1^3 b_0 \psi_1^4 X_0^2} + \frac{3 a_2 b_1}{a_1^3 b_0^2 \psi_1^3 X_0^2}$$

$$\quad + \frac{3 \psi_2^2}{a_1 b_0 \psi_1^5 X_0} + \frac{6 b_1 \psi_2}{a_1 b_0^2 \psi_1^4 X_0} - \frac{\psi_3}{a_1 b_0 \psi_1^4 X_0}$$

$$\quad + \frac{6 b_1^2}{a_1 b_0^3 \psi_1^3 X_0} - \frac{3 b_2}{a_1 b_0^2 \psi_1^3 X_0} - \frac{a_3}{a_1^4 b_0 \psi_1^3 X_0^3}.$$

Here, $a_j, b_j, \varphi_j, \psi_j$ denote the jth Taylor coefficients of the functions a, b, φ, ψ, respectively.

27. Equivalent version of reciprocity [p. 117]. We will use Eq. (8.5) from Remark 8.2 to show that the reciprocity law (8.4) and the main result from Theorem 2.3 (= Theorem 6.1 in [84]) are equivalent statements.

Let us start with the following equivalent formulation of Theorem 2.3 (obtained by simply shifting the summation index):

$$A_{n,k} = \sum_{j=0}^{n-k} (-1)^j \binom{n-1+j}{k-1} X_1^{-n-j} \widetilde{B}_{n-k+j,j}. \tag{1}$$

In (1), replace k by $n-k$ and apply the reciprocity of the binomial coefficients (8.1). This leads to

$$A_{n,n-k} = (-1)^k \sum_{j=0}^{k} \binom{-n+k}{k+j} X_1^{-n-j} \widetilde{B}_{k+j,j}. \tag{2}$$

Now, by taking $-n+k$ for n in Eq. (8.5) we obtain

$$B_{-n+k,-n} = \sum_{j=0}^{k} \binom{-n+k}{k+j} X_1^{-n-j} \widetilde{B}_{k+j,j}, . \tag{3}$$

(2) and (3) imply the following equivalent formulation of Eq. (8.4):

$$B_{-n+k,-n} = (-1)^k A_{n,n-k}. \tag{4}$$

Obviously (1) $\iff$ (2) $\iff$ (4). ◊

It only takes a little more to verify that also the Generalized Schlömilch-Schläfli identities (Theorem 5.5) and the General Reciprocity Law (Theorem 8.1) are equivalent statements.

Final Remark. — In Section 8 the reciprocity law was proved without using Theorem 6.1 (from Chapter I). Since, in the reverse direction, this main result can be derived from the reciprocity law, we now have an independent second proof that is not based on an inductive argument.

APPENDIX: A MATHEMATICA PACKAGE

In connection with my research on multivariate Stirling polynomials (MSP), I developed a package for the computer algebra system *Mathematica*® (Version 7 or higher)[1]. It offers functions that generate these polynomials; a simple substitution mechanism is also available in the package's instruction set. For more details cf. [82].

Instruction set

`MultivariateStirlingP1[n,k]`
$\rightarrowtail$ Gives the polynomial $S_{n,k} \in \mathbb{Z}[X_1, \ldots, X_{n-k+1}]$ defined by $X_1^{2n-1} A_{n,k}$

`MultivariateStirlingA[n,k]`
$\rightarrowtail$ Gives the (Laurent) polynomial $A_{n,k}$ (MSP of the 1st kind)

`MultivariateStirlingP2[n,k]`
$\rightarrowtail$ Gives the polynomial $B_{n,k} \in \mathbb{Z}[X_1, \ldots, X_{n-k+1}]$ (MSP of the 2nd kind)

`AssociateBellPolynomial[n,k]`
$\rightarrowtail$ Gives the associated Bell polynomial $B_{n,k}(0, X_2, \ldots, X_{n-k+1})$

`CauchyPolynomial[n,k]`
$\rightarrowtail$ Gives the (partial) cycle indicator polynomial $Z_{n,k} \in \mathbb{Z}[X_1, \ldots, X_{n-k+1}]$

`SetVariablesTo[{var1, var2, ...}]`
$\rightarrowtail$ A rule that replaces `X[1]`, `X[2]`, ... by $var1, var2, \ldots$

`SubIndexed[m]`
$\rightarrowtail$ A rule that replaces `X[1]`, ..., `X[m]` by $X_1, \ldots, X_m$

How to use the package

Remark. The symbol X is used as the basis letter denoting indeterminates; it is protected within this package, that is, you cannot change its value.

Nevertheless it is possible to substitute any values into X_j $(j = 1, 2, \ldots)$:

```
In: {X[1],X[2],X[3]}/.SetVariablesTo[{-5,7}]
```

---

[1] *Mathematica* is a registered trademark of Wolfram Research, Inc.

```
Out: {-5,7,X[3]}
```

Read in the package file. In order to evaluate the cell below, the file MultivariateStirlingPolynomials.m must have been copied[9] into your working directory:

```
SetDirectory[NotebookDirectory[]];
<< MultivariateStirlingPolynomials'
```

Generating multivariate Stirling polynomials. MSPs of the second kind are the same as partial Bell Polynomials.

Let us, for example, generate $B_{6,4}$:

```
In: MultivariateStirlingP2[6,4]
Out: 45 X[1]^2 X[2]^2 + 20 X[1]^3 X[3]
```

If you do not like the indeterminates notated as X[1],X[2],X[3],..., then try

```
MultivariateStirlingP2[6,4]/.SetVariablesTo[{x,y,z}]
```
$45x^2y^2 + 20x^3z$

or

```
MultivariateStirlingP2[6,4]/.SubIndexed[3]
```
$45X_1^2X_2^2 + 20X_1^3X_3$

Unification, that is, the replacement of all indeterminates by 1, gives the sum of the coefficients (a Stirling number of the second kind):

```
MultivariateStirlingP2[6,4]/.SetVariablesTo[{1,1,1}]
65
```

Let us now create a triangular matrix of partial Bell polynomials:

```
In: BMatrix = Table[Table[
    MultivariateStirlingP2[i,j],{j,1,4}],{i,1,4}];
    BMatrix /. SubIndexed[4] // MatrixForm
Out:
```

$$\begin{pmatrix} X_1 & 0 & 0 & 0 \\ X_2 & X_1^2 & 0 & 0 \\ X_3 & 3X_1X_2 & X_1^3 & 0 \\ X_4 & 3X_2^2 + 4X_1X_3 & 6X_1^2X_2 & X_1^4 \end{pmatrix}$$

---

[9] The package file is available at the URL given in [82].

Similarly we obtain the matrix of the orthogonal companions $A_{n,k}$:

```
In: AMatrix = Table[Table[
      MultivariateStirlingA[i,j],{j,1,4}],{i,1,4}]
      // MatrixForm
Out:
```

$$
\begin{pmatrix}
\frac{1}{X_1} & 0 & 0 & 0 \\
-\frac{X_2}{X_1^3} & \frac{1}{X_1^2} & 0 & 0 \\
\frac{3X_2^2}{X_1^5} - \frac{X_3}{X_1^4} & -\frac{3X_2}{X_1^4} & \frac{1}{X_1^3} & 0 \\
-\frac{15X_2^3}{X_1^7} + \frac{10X_2X_3}{X_1^6} - \frac{X_4}{X_1^5} & \frac{15X_2^2}{X_1^6} - \frac{4X_3}{X_1^5} & -\frac{6X_2}{X_1^5} & \frac{1}{X_1^4}
\end{pmatrix}
$$

The fundamental inversion law for the MSPs then looks like this:

```
In: AMatrix.BMatrix // Expand // MatrixForm
```

$$
\begin{pmatrix}
1 & 0 & 0 & 0 \\
0 & 1 & 0 & 0 \\
0 & 0 & 1 & 0 \\
0 & 0 & 0 & 1
\end{pmatrix}
$$

Source code of the package

Auxiliary functions. Functions in the Private part of the implementation:

```
PartitionSet[n_Integer,k_Integer]:=
   Select[Partitions[n,n],Length[#]==k&]
```

(Generates the set of all (n, k)-partitions)

```
PartitionListToType[p_List]:=Module[
   {ptf={},pset=Union[p]},
   Do[AppendTo[ptf,{pset[[i]],Count[p,pset[[i]]]}],
   {i,1,Length[pset]}];ptf]
```

(Gives the type of a partition p)

```
NumberOfElements[s_List]:=Module[
    {t1=Table[{i,1},{i,1,Length[s]}],
    t2=Table[{i,2},{i,1,Length[s]}]},
    Extract[s,t1].Extract[s,t2]]
```

(Gives the number $r_1 + 2r_2 + 3r_3 + \cdots$ of elements of partition type s)

```
NumberOfBlocks[s_List]:=
    Plus@@Extract[s,Table[{i,2},{i,1,Length[s]}]]
```

(Gives the number $r_1 + r_2 + r_3 + \cdots$ of blocks of the partition type s)

```
NumberOfBlockPermutations[s_List]:=
    Module[{rtab=Extract[s,Table[{i,2},{i,1,Length[s]}]]},
    Times@@Factorial[rtab]]
```

(Gives the number $r_1!r_2!r_3!\cdots$ of permutations of the blocks of s)

```
NumberOfElementPermutations[s_List]:=Module[
    {t1=Table[{i,1},{i,1,Length[s]}],
    t2=Table[{i,2},{i,1,Length[s]}]},
    Times@@(Factorial[Extract[s,t1]]^Extract[s,t2])]
```

(Gives the number $(1!)^{r_1}(2!)^{r_2}(3!)^{r_3}\cdots$ of permutations of the elements in the unordered blocks)

```
StirlingFunction[ps_List]:=Module[
    {r1, k, n},
    If[Cases[ps,{1,_}]=={},r1=0,r1=Extract[ps,{1,2}]];
    n=NumberOfBlocks[ps]+1;
    k=2*n-1-NumberOfElements[ps];
    (-1)^(n-1-r1)*(2*n-2-r1)! *
    r1!/((k-1)!*NumberOfBlockPermutations[ps] *
    NumberOfElementPermutations[ps])]
```

(Gives the coefficients of the Stirling polynomials of type P1)

```
SubsetFunction[ps_List]:=
    Factorial[NumberOfElements[ps]]/
    (NumberOfBlockPermutations[ps] *
    NumberOfElementPermutations[ps])
```

(Gives the coefficients of the Stirling polynomials of Bell type P2)

```
PartitionToMonomial[ps_List]:=Module[
   {n=NumberOfElements[ps],XList},
   XList=ps/.{j_Integer,r_Integer}:>(Array[X,n][[j]])^r;
   Times@@XList]
```

(Converts a partition type into a monomial with indeterminates
X[1],X[2],...)

Functions intended for export. The functions listed below correspond to the
set of commands introduced at the beginning. These are the functions the
user can call in his *Mathematica* environment. Since the meaning of the
symbols has already been explained at the beginning, I may confine myself
to specifying their source code here.

```
MultivariateStirlingP1[n_Integer,k_Integer]:=Module[
   {s=Map[PartitionListToType,PartitionSet[2*n-1-k,n-1]]},
   Map[StirlingFunction,s].Map[PartitionToMonomial,s]]

MultivariateStirlingA[n_Integer,k_Integer]:=
   Expand[MultivariateStirlingP1[n,k]/X[1]^(2*n-1)]

MultivariateStirlingP2[n_Integer,k_Integer]:=Module[
   {pss=Map[PartitionListToType,PartitionSet[n,k]]},
   Map[SubsetFunction,pss].Map[PartitionToMonomial,pss]]

SetVariablesTo[vars_List]:=Module[
   {xtab},
   Unprotect[X];
   xtab=Table[X[j] -> vars[[j]],{j,1,Length[vars]}];
   Protect[X];
   xtab]

SubIndexed[m_Integer]:=
   SetVariablesTo[Table[Subscript[X,j],{j,1,m}]]

AssociateBellPolynomial[n_Integer,k_Integer]:=
   MultivariateStirlingP2[n,k]/.SetVariablesTo[{0}]

CauchyPolynomial[n_Integer,k_Integer]:=
   MultivariateStirlingP2[n,k] /.
   SetVariablesTo[Table[(j-1)!X[j],{j,1,n-k+1}]]
```

BIBLIOGRAPHY

[1] M. ABBAS and S. BOUROUBI: On new identities for Bell's polynomials. *Discrete Math.* **293** (2005), 5–10.

[2] U. ABEL: New proofs of Melzak's identity. *Aequationes mathematicae* **94** (2020), 163–167.

[3] M. AIGNER: *Combinatorial Theory* (Grundlehren der mathematischen Wissenschaften, 234), Springer-Verlag: Berlin – New York (1979).

[4] T. M. APOSTOL: Calculating higher derivatives of inverses. *Amer. Math. Monthly* **107** (2000), 738–741.

[5] L. F. A. ARBOGAST: *Du calcul des dérivations.* Strasbourg (1800).

[6] E. T. BELL: Exponential polynomials. *Annals of Mathematics* **35** (1934), 258–277.

[7] P. BENDER: Eine Bemerkung zu Derivationen von Potenzreihen. *Mathematisch-physikalische Semesterberichte* **27**/1 (1980), 143–145.

[8] F. BERGERON and C. REUTENAUER; Une interprétation combinatoire des puissances d'un opérateur differentiel linéaire. *Ann. Sci. Math. Qué.* **11** (1987), 269–278.

[9] J. BERTRAND: *Traité de calcul différentiel et de calcul intégral (Première partie – Calcul différentiel).* Gauthier-Villars, Paris (1864).

[10] D. BIRMAJER, J. B. GIL, and M. D. WEINER: Some convolution identities and an inverse relation involving partial Bell polynomials. *The Electronic J. of Comb.* **19** (2012), no. 4, Paper 34.

[11] U. T. BÖDEWALDT: Die Kettenregel für höhere Ableitungen. *Math. Z.* **48** (1942), 735–746.

[12] K. N. BOYADZHIEV: Power sum identities with generalized Stirling numbers. *Fibonacci Quarterly* **46/47** (2009), 326–330.

[13] ———: Close Encounters with the Stirling Numbers of the Second Kind. *Math. Mag* **85** (2012), 252–266.

[14] _____: Lah numbers, Laguerre polynomials of order negative one, and the nth derivative of $\exp(1/x)$. *Acta Univ. Sapientiae, Mathematica* **8**/1 (2016), 22–31.

[15] _____: Melzak's formula for arbitrary polynomials. *Util. Math.* **99** (2016), 397–401.

[16] CH. BROUDER, A. FRABETTI, and CH. KRATTENTHALER: Non-commutative Hopf algebra of formal diffeomorphisms. *Adv. Math.* **200**/2 (2006), 479–524.

[17] R. P. BRENT AND H. T. KUNG: Fast algorithms for manipulating formal power series. *J. Assoc. Comput. Mach* **25** (1978), 581–595.

[18] CH. A. CHARALAMBIDES: *Enumerative Combinatorics*. Chapman & Hall/CRC, Boca Raton 2002.

[19] W.-S. CHOU, L. C. HSU, and P. J.-S. SHIUE: Applications of Faà di Bruno's formula in characterization of inverse relations. *J. Comp. and Appl. Math.* **190** (2006), 151–169.

[20] A. CONNES AND D. KREIMER: Hopf algebras, renormalization and non-commutative geometry. *Commun. Math. Phys.* **199** (1998), 203–242; hep-th/9808042.

[21] L. COMTET: Une formule explicite pour les puissances successives de l'opérateur de dérivations de Lie. *C. R. Acad. Sci.* **276** (1973), 165–168.

[22] _____: *Advanced Combinatorics*, rev. and enlarged edition. Reidel, Dordrecht (Holland) 1974.

[23] A. D. D. CRAIK: Prehistory of Faà di Bruno's Formula. *Amer. Math. Monthly* **112** (2005), 119–130.

[24] D. CVIJOVIĆ: New identities for the partial Bell polynomials. *Appl. Math. Lett.* **24**/9 (2011), 1544–1547.

[25] G. DOBIŃSKI: Summirung der Reihe $\sum n^m/n!$ für $m = 1, 2, 3, 4, 5, \ldots$ *Grunert's Archiv* **61** (1877), 333–336.

[26] W. A. DUDEK and V. S. TROKHIMENKO: *Algebras of Multiplace Functions*. De Gruyter, Berlin/Boston (2012).

[27] F. QI, D.-W. NIU, D. LIM and Y.-H. YAO: Special values of the Bell polynomials of the second kind for some sequences and functions. *Journal of Mathematical Analysis and Applications* **491**/2 (July 2020). DOI: 10.1016/j.jmaa.2020.124382

[28] H. FIGUEROA and J. M. GRACIA-BONDÍA: Combinatorial Hopf algebras in quantum field theory I. *Review in Mathematical Physics* **17** (8) (2005), 881–976.

[29] D. FINKELSHTEIN, Y. KONDRATIEV, E. LYTVYNOV, and M. J. OLIVEIRA: Stirling operators in spatial combinatorics. Preprint: arXiv: 2007.01175v4 (18 Aug 2020).

[30] I. GESSEL: Lagrange inversion. *J. Comb. Theory (A)* **144** (2016), 212–249.

[31] I. GESSEL and R. P. STANLEY: Stirling polynomials. *J. Comb. Theory (A)* **24** (1978), 24–33.

[32] H. W. GOULD: Stirling number representation problems. *Proc. Amer. Math. Soc.* **11** (1960), 447–451.

[33] M. HAIMAN and W. R. SCHMITT: Incidence algebra antipodes and Lagrange inversion in one and several variables. *J. Comb. Theory (A)* **50** (1989), 172–185.

[34] B. HARRIS and L. SCHOENFELD: The number of idempotent elements in symmetric semigroups. *J. Comb. Theory* **3**/2 (1967), 122–135.

[35] H. HAN and S. SEO: Combinatorial proofs of inverse relations and log-concavity for Bessel numbers. *European J. Comb.* **29**/7 (2008), 1544–1554.

[36] F. T. HOWARD: Associated Stirling numbers. *The Fibonacci Quarterly* **18** (1980), 303–315.

[37] L. C. HSU: A summation rule using Stirling numbers of the second kind. *Fibonacci Quarterly* **31** (1993), 256–262.

[38] I-CH. HUANG: Inverse relations and Schauder bases. *J. Comb. Theory (A)* **97** (2002), 203–224.

[39] E. JABOTINSKY: Sur la réprésentation de la composition des fonctions par un produit de matrices. Application à l'itération de e^x et de $e^x - 1$. *C. R. Acad. Sci.* **224** (1947), 323–324.

[40] ———: Sur les fonctions inverses. *C. R. Acad. Sci.* **229** (1949), 508–509.

[41] ———: Representation of functions by matrices. Application to Faber polynomials. *Proc. Amer. Math. Soc.* **4** (1953), 546–553.

[42] W. P. JOHNSON: The curious history of Faà di Bruno's formula. *Amer. Math. Monthly* **109** (2002), 217–234.

[43] ———: Combinatorics of higher derivatives of inverses. *Amer. Math. Monthly* **109** (2002), 273–277.

[44] C. JORDAN: *Calculus of Finite Differences*, 1st ed. Budapest (1939); 2nd ed., Repr., Chelsea Publ. Co., Inc., New York (1950).

[45] F. KAMBER: Formules exprimant les valeurs des coefficients des séries de puissances inverses. *Acta Math.* **78** (1946), 193–204.

[46] L. KARGIN: Some formulæ for products of geometric polynomials with applications. *J. of Integer Sequences* **20** (2017), Article 17.4.4.

[47] S. KHELIFA and Y. CHERRUAULT: Nouvelle identité pour les polynômes de Bell. *Maghreb Math. Rev.* **9** (2000), 115–123.

[48] P. M. KNOPF: The operator $(x\frac{d}{dx})^n$ and its application to series. *Math. Mag.* **76** (2003), 364–371.

[49] D. E. KNUTH: Convolution polynomials. *Math. J.* **2** (1992), 67–78.

[50] ———: Two notes on notation. *Amer. Math. Monthly* **99** (1992), 403–422.

[51] ———: *The Art of Computer Programming*, 3rd ed., Vol. 1: *Fundamental Algorithms*. Addison Wesley Longman (1997).

[52] D. E. KNUTH and B. PITTEL: A recurrence related to trees. *Proc. Amer. Math. Soc.* **105**/2 (1989), 335–349.

[53] E. R. KOLCHIN: *Differential algebra and algebraic groups*. New York and London, 1973.

[54] CH. KRATTENTHALER: Operator methods and Lagrange inversion: A unified approach to Lagrange formulas. *Trans. Amer. Math. Soc.* **305** (1988), 431–465.

[55] D. Kreimer: On the Hopf algebra structure of perturbative field theory. *Adv. Theor. Math. Phys.* **2.2** (1998), 303–334; q-alg/9707029.

[56] J. L. Lagrange: Nouvelle méthode pour résoudre les équations littérales par le moyen des séries. *Mém. Acad. Roy. Sci. Belles-Lettres de Berlin* **24** (1770), 251–326 [also in Œuvres de Lagrange, vol. 3, Gauthier-Villars, Paris (1869), pp. 5–73].

[57] I. Lah: Eine neue Art von Zahlen, ihre Eigenschaften und Anwendung in der mathematischen Statistik. *Mitt.-Bl. Math. Statistik.* **7** (1955), 203–212.

[58] S. K. Lando: *Lectures on Generating Functions.* Student Mathematical Library, Volume 23, The American Mathematical Society, 2003.

[59] T. Mansour and M. Schork: *Commutation Relations, Normal Ordering, and Stirling Numbers.* CRC Press, Boca Raton (2016).

[60] P. J. McCarthy: Functional nth roots of unity. *The Mathematical Gazette* **64** (1980), 107–115.

[61] M. A. McKiernan: On the nth derivative of composite functions. *Amer. Math. Monthly* **63** (1965), 331–333.

[62] Z. A. Melzak: Problem 4458. *Amer. Math. Monthly* **58** (1951), 636.

[63] Z. A. Melzak, V. D. Gokhale, R. V. Parker: Advanced problems and solutions. Solutions: 4458. *Amer. Math. Monthly* **60** (1953), 53–54.

[64] K. Menger: *Algebra of Analysis*, no. 3 in *Notre Dame Mathematical Lectures.* Notre Dame, Indiana (1944).

[65] _____: The algebra of functions: past, present, future. *Rend. Mat. Appl.*, V. Ser. 20 (1961), 409–430.

[66] M. Mihoubi: *Polynômes multivariés de Bell et polynômes de type binomial.* Thèse de Doctorat, Université de Sciences et de la Technologie Houari Boumedienne (2008).

[67] _____: Bell polynomials and binomial type sequences. *Discrete Math.* **308** (2008), 2450–2459.

[68] _____: Partial Bell polynomials and inverse relations. *J. of Integer Sequences* **13** (2010), Article 10.4.5.

[69] Ž. MIJAJLOVIĆ and Z. MARKOVIĆ: Some recurrence formulas related to the differential operator θd. *Facta Universitatis* (NIŠ) **13** (1998), 7–17.

[70] S. C. MILNE and G. BHATNAGAR: A characterization of inverse relations. *Discrete Math.* **193** (1998), 235–245.

[71] P. M. MORSE and H. FESHBACH: *Methods of Theoretical Physics*, Vol. I. McGraw-Hill Book Co., Inc., New York (1953).

[72] R. MULLIN and G.-C. ROTA: On the foundations of combinatorial theory III: Theory of binomial enumeration. In: *Graph theory and its applications*, Academic Press, Inc., San Diego 1970.

[73] N. NIELSEN: Recherches sur les polynômes et les nombres de Stirling. *Annali di Matematica pura ed applicata (series 3)* **10** (1904), 287–318.

[74] _____: Über die Stirlingschen Polynome und die Gammafunktion. *Monatshefte für Mathematik und Physik* **16** (1) (1905), 135–140.

[75] _____: *Traité élémentaire des nombres de Bernoulli*. Paris 1923.

[76] J. QUAINTANCE and H. W. GOULD: *Combinatorial Identities for Stirling Numbers*. The Unpublished Notes of H. W. Gould. World Scientific 2016.

[77] J. RIORDAN: *Introduction to Combinatorial Analysis*. New York, 1958, reprint Dover Publ., Inc., Mineola, New York 2002.

[78] _____: *Combinatorial Identities*. New York 1968.

[79] O. SCHLÖMILCH: Recherches sur le coéfficients des facultés analytiques. *Crelle's Journal für die reine und angewandte Mathematik* **44** (1852), 344–355.

[80] W. R. SCHMITT: *Antipodes and Incidence Coalgebras*, Ph. D. thesis, MIT, 1986.

[81] _____: Incidence Hopf Algebras. *J. Pure Appl. Algebra* **96** (3) (1994), 299–330.

[82] A. Schreiber: A *Mathematica* Package for the Multivariate Stirling Polynomials (Nov 2013). http://www.gefilde.de/ashome/software /msp/stirling.html

[83] ——: *Didaktische Schriften zur Elementarmathematik*. Logos Verlag, Berlin 2014.

[84] ——: Multivariate Stirling polynomials of the first and second kind. *Discrete Math.* **338** (2015), 2462–2484. — Preprint: arXiv:1311.5067.

[85] ——: Eine Dobiński-Reihe für die Anzahl surjektiver Abbildungen. *Elem. Math.* **73** (2018), 130–132.

[86] ——: Inverse relations and reciprocity laws involving partial Bell polynomials and related extensions. *Enumer. Combin. Appl.* **1**:1 (2021), Article S2R3. — Preprint: arXiv:2009.09201.

[87] E. Schröder: Vier combinatorische Probleme. *Ztschr. für Math. Phys.* **15** (1870), 361–376.

[88] I. Schur: On Faber polynomials. *Amer. J. Math.* **67** (1945), 33–41.

[89] B. Schweizer and A. Sklar: The algebra of functions I–III. *Math. Ann.* **139** (1960), 366–382; **143** (1961), 440–447; **161** (1965), 171–196.

[90] ——: Function systems. *Math. Ann.* **172** (1967), 1–16.

[91] G. Scott: Formulae of successive differentiation. *Quarterly J. Pure Appl. Math.* **4** (1861), 77–92.

[92] A. Sklar: Commentary on the Algebra of Analysis and Algebra of Functions, in *Karl Menger Selecta Mathematica*, vol. 2 (ed. by Bert Schweizer, Abe Sklar et. al.), Springer, Wien 2003, 111–126.

[93] R. P. Stanley: Combinatorial reciprocity theorems. *Adv. Math.* **14** (1974), 194–253.

[94] ——: *Enumerative Combinatorics*, Vol. 1, Wadsworth & Brooks/Cole, Monterey 1986.

[95] ——: *Enumerative Combinatorics*, Vol. 2, Cambridge University Press 1999.

[96] I. Suciu: *Function Algebras.* Noordhoff International Publishing, 1975.

[97] J. J. Sylvester: Note on Burman's Law for the inversion of the independent variable. *Philosophical Magazine*, 4th series **8** (1854), 535–540. — *The Collected Mathematical Papers of James Joseph Sylvester* II, Cambridge 1908, pp. 44–49.

[98] _____: On reciprocants. *Messenger of Mathematics* **15** (1886), 88–92. — *The Collected Mathematical Papers of James Joseph Sylvester* IV, Cambridge 1912, pp. 255–258.

[99] P. G. Todorov: New explicit formulas for the n-th derivative of composite functions (german). *Pacific Journ. of Math.* **92**/1 (1981), 217–236.

[100] _____: New differential recurrence relations for Bell polynomials and Lie coefficients. *Comptes rendus de l'Académie bulgare des Sciences* **38**/1 (1985), 43–45.

[101] L. Toscano: Sulla iterazione dell'operatore xd. *Rendiconti di matematica e delle sui applicazioni* **V** (8) (1949), 337–350.

[102] W. Wang and T. Wang: General identities on Bell polynomials. *Computers & Mathematics with Applications* **58** (2009), no. 1, 104–118.

[103] _____: Matrices related to the idempotent numbers and the numbers of planted forests. *Ars Combin.* **98** (2011), 83–96.

[104] E. T. Whittaker: On the reversion of series. *Gaz. Mat. Lisboa* **12** (50) (1951), 1.

[105] Sh.-l. Yang: Some identities involving the binomial sequences. *Discrete Math.* **308** (2008), 51–58.

Acknowledgement

Parts of Chapter I are reprinted from *Discrete Mathematics*, **338**/12, Alfred Schreiber: Multivariate Stirling Polynomials of the first and second Kind, pp. 2462–2484, Copyright 2015, with permission from Elsevier.

OTHER BOOKS BY THE AUTHOR

Werktage im Niemandsland
Aus dem Fahrtenbuch eines mathematischen Grenzgängers
Logos-Verlag: Berlin 2017 (245 pp.)
ISBN 978-3-8325-4379-2

Didaktische Schriften zur Elementarmathematik
Logos-Verlag: Berlin 2014 (255 pp.)
ISBN 978-3-8325-3716-6

Die enttäuschte Erkenntnis
Paramathematische Denkzettel.
Edition am Gutenbergplatz: Leipzig 2013 (211 pp.)
ISBN 978-3-937219-68-4

Begriffsbestimmungen
Aufsätze zur Heuristik und Logik mathematischer Begriffsbildung.
Logos-Verlag: Berlin 2011 (308 pp.)
ISBN 978-3-8325-2883-6

Operative Genese der Geometrie
(mit Peter Bender) hpt & Teubner: Wien / Stuttgart 1985.
Reprint bei Neopubli: Berlin 2012 (468 pp.)
ISBN 978-3-8442-2454-2

Theorie und Rechtfertigung
Untersuchungen zum Rechtfertigungsproblem axiomatischer Theorien.
Friedr. Vieweg & Sohn: Braunschweig 1975 (204 pp.)
ISBN 978-3-528-08345-8